THE AI INTERVIEWER
PRODUCT MANAGER EDITION
101 Questions and AI-generated solutions

RUSHAD HEERJEE

Copyright © 2023 Rushad Kaizad Heerjee

All rights reserved. No part of this book may be reproduced, stored in a retrieval system, or transmitted in any form or by any means, electronic, mechanical, photocopying, recording, or otherwise, without the prior written permission of the copyright owner.

The information contained in this book is based on the research and experience of the author. The author has made every effort to provide accurate and up-to-date information. However, the author and publisher do not assume any liability for errors or omissions. The author and publisher shall not be held responsible for any damages resulting from the use of the information contained in this book.

Cover design by Moiz Najmi
ISBN: 9798374169423
Imprint: Independently published
First edition 2023
This book was printed in the United States of America

Contents

Preface v

Chapter 1 Introduction 1
   Why is the PM interview important? 1
   What are common mistakes people make during the PM interview? 2
   How can this book help you prepare? 5

Chapter 2 Understanding the PM Role 7
   What is a PM and what do they do? 7
   What is the difference between a product and program manager? 10
   What skills and qualities do successful PMs have? 12
   How can you demonstrate these skills and qualities during the interview? 13

Chapter 3 Top PM Interview Questions 17
   An overview of the types of questions that may be asked during the PM interview 17
   The STAR Method 20
   101 top PM interview questions 22

Chapter 4 Behavioral Questions Answers and Explanations 33

Chapter 5 Role-Specific Questions Answers and Explanations 63

Chapter 6 Case Questions Answers and Explanations 99

Chapter 7 Obscure Questions Answers and Explanations 123

Chapter 8 Technical (Code) Based Questions Answers and Explanations 139

Chapter 9 Scenario Based Technical (Code) Questions Answers and Explanations 155

Chapter 10 Brainteaser Questions Answers and Explanations 169

Chapter 11 Encouragement 175

Preface

Product management is a challenging and rewarding field that requires a unique blend of technical, business, and leadership skills. Whether you are an aspiring product manager looking to break into the industry, or an experienced product leader looking to advance your career, an impressive performance in your product management interviews is critical to achieving your goals.

As you embark on your product management job search, you may be wondering what to expect in your interviews and how to stand out from the competition. You may be asking yourself questions like: What kind of questions will I be asked? How can I demonstrate my skills and fit for the role? How can I effectively communicate my vision and approach to product development?

These are all important considerations, and this book is here to help. Inside, you will find 101 product management interview questions, along with AI-generated example solutions to help you understand how to approach and answer each one. These questions cover a wide range of topics that are commonly tested in product management interviews, including:

- *Strategy and vision*: How do you set product vision and roadmap? How do you prioritize features and evaluate business impact? How do you assess market trends and competitive landscape?
- *Product development*: How do you define and scope products? How do you gather and synthesize customer feedback? How do

you design and test solutions? How do you manage cross-functional teams and stakeholders?

- *Analytics and data*: How do you measure and track product performance? How do you use data to inform decision-making? How do you define and track key metrics and targets?

- *Leadership and communication*: How do you lead and motivate teams? How do you communicate effectively with different audiences? How do you negotiate and make decisions under uncertainty?

By working through these questions and example solutions, you will gain a deeper understanding of the skills and competencies that are essential for product management success. You will also have the opportunity to practice your interview skills and develop your own responses and examples based on your own experiences and insights.

Of course, it is important to note that these questions and examples are meant to serve as a guide and are not exhaustive. Every company and role are different, and you should always tailor your responses to the specific context and requirements of the position you are applying for.

With that said, we hope this book will provide you with a strong foundation and confidence to succeed in your product management interviews and land that dream job at a technology company. We wish you the best of luck in your job search, and we hope you will find this book to be a valuable resource in your journey to becoming a product management leader.

– Rushad Heerjee

CHAPTER 1

Introduction

WHY IS THE PM INTERVIEW IMPORTANT?

The product manager interview is a crucial step in the hiring process for product management roles. It is an opportunity for both the candidate and the company to assess fit and determine if the candidate has the skills and experience necessary to succeed in the role.

As a product manager, you will be responsible for leading the development and success of a product. This involves setting the product vision and roadmap, gathering and synthesizing customer feedback, defining and tracking key metrics, and leading cross-functional teams to deliver high-quality products on time and within budget. It also requires strong analytical, communication, and leadership skills.

Therefore, it is important for both the candidate and the company to have a clear understanding of the expectations and requirements of the product manager role during the interview process. The product manager interview is the time to highlight your skills and experience, and to demonstrate your fit for the role and the company culture.

There are a wide variety of questions that may be asked in a product manager interview, covering topics such as strategy and vision, product development, analytics and data, and leadership and communication. Some common product manager interview questions include:

- How do you set product vision and roadmap?
- How do you prioritize features and evaluate business impact?
- How do you gather and synthesize customer feedback?
- How do you measure and track product performance?
- How do you lead and motivate teams?
- How do you communicate effectively with different audiences?

Answering these questions effectively requires a sturdy foundation in product management concepts and skills, as well as the ability to provide specific examples from your past experiences. It is also important to be able to tailor your responses to the specific needs and challenges of the company and role you are interviewing for.

WHAT ARE COMMON MISTAKES PEOPLE MAKE DURING THE PM INTERVIEW?

The product management interview process at top technology companies is a critical step in the hiring process, and it is important to be well-prepared and confident when you sit down for your interview. These companies are known for their innovative and data-driven approaches to product development, and they expect candidates to have strong technical, business, and leadership skills. As such, it is important to avoid common mistakes during the product management interview.

One common mistake is failing to research the company and the product manager role. It is important to familiarize yourself with the company's products, vision, and culture, as well as the specific requirements and responsibilities of the product manager role for whom you are applying. This will enable you to tailor your responses and examples to the specific context of the interview, and to demonstrate your fit for the role and the company. This includes understanding the company's target market and customers, their business model and revenue streams, their competitive landscape, and their culture and values.

Another common mistake is being unable to clearly articulate your product vision and strategy. As a product manager, you will be responsible for setting the direction and roadmap for your product, and it is important to be able to clearly communicate your vision and approach to the interviewers. This requires a deep understanding of the customer needs and market trends, as well as the ability to prioritize and make trade-offs based on data and business impact. It is also important to be able to articulate the value proposition and unique selling points of your product, and to demonstrate how it fits into the company's overall product strategy.

A third common mistake is being unable to provide specific examples of your past experiences and achievements. Product management interviews often involve behavioral questions that ask you to describe how you have approached and solved problems in the past. It is important to have a set of well-crafted examples that highlight your skills and achievements, and to be able to provide concrete details and results. These examples should highlight your ability to gather and synthesize customer feedback, define, and track key metrics, make data-driven decisions, and lead cross-functional teams to deliver high-quality products on time and within budget.

A fourth common mistake is being unprepared for technical questions. While product management roles do not typically require deep technical expertise, it is important to have a basic understanding of core technical concepts, as well as the ability to learn and adapt quickly. This may include understanding the technology stack and architecture of the company's products, as well as how different technologies and platforms fit together. It is also important to be able to communicate effectively with technical stakeholders and to understand the trade-offs and constraints involved in product development.

A fifth common mistake is failing to display your leadership and communication skills. Product management involves leading and motivating cross-functional teams, as well as communicating effectively with different audiences. It is important to be able to demonstrate your leadership style and communication skills through specific examples and anecdotes. This includes the ability to listen and empathize with customers, to inspire and empower your team, to negotiate and make decisions under uncertainty, and to present and defend your ideas with clarity and conviction.

The product management interview process at top technology companies is a crucial step in the hiring process, and it is important to be well-prepared and confident when you sit down for your interview. By demonstrating your skills and fit for the role, you can increase your chances of landing that dream job as a product manager at a technology company. It is also important to be aware of common mistakes and to take steps to avoid them to maximize your chances of success.

HOW CAN THIS BOOK HELP YOU PREPARE?

As an aspiring product manager, you may be wondering how to best prepare for and succeed in your product management interviews. One effective way to do this is by practicing with a range of product management interview questions and working through example solutions.

This book is focused on helping people master the product management interview through questions and answers is a valuable resource for anyone looking to break into the field or advance their career in product management. It can help you:

- Familiarize yourself with the types of questions that are commonly asked in product management interviews
- Understand how to approach and answer each question, including the key skills and competencies that are being tested
- Develop your own responses and examples based on your own experiences and insights
- Practice your interview skills and build confidence in your abilities

This book features a collection of product management interview questions and answers that have been gathered from some of the finest PM hiring managers in the industry. These questions cover a wide range of topics, including strategy and vision, product development, analytics and data, and leadership and communication.

By working through the questions and example solutions in this book, you will gain a deeper understanding of the skills and competencies that are essential for product management success. You will also have the opportunity to practice your interview skills and develop your own responses and examples based on your own experiences and insights.

Of course, it is important to note that the questions and examples in the book are meant to serve as a guide and are not exhaustive. Every company and role are different, and you should always tailor your responses to the specific context and requirements of the position you are applying for.

In summary, the materials in this book are focused on helping people master the product management interview through questions and answers is a crucial resource for anyone looking to break into the field or advance their career in product management. By familiarizing yourself with the types of questions that are commonly asked, understanding how to approach and answer them, and practicing your interview skills, you can increase your chances of landing that dream job as a product manager at a technology company.

CHAPTER 2

Understanding the PM Role

WHAT IS A PM AND WHAT DO THEY DO?

Product management is a critical role in the technology industry, responsible for defining, building, and delivering high-quality products that meet the needs of customers and drive business growth. The product manager is the key leader and decision-maker for their product, and they play a vital role in shaping the direction and success of the product and the company.

So, what is a product manager and what do they do?

A product manager is a strategic leader who is responsible for the overall direction, vision, and success of a product. They work closely with cross-functional teams, including engineering, design, marketing, sales, and support, to define the product strategy, roadmap, and feature set. They also work with customers and stakeholders to gather and analyze market and user feedback, and to make data-driven decisions based on customer needs, market trends, and business impact. The product manager is the owner of the product, and they have a wide range of responsibilities that can vary depending on the specific company and industry.

The product manager is the owner of the product, and they have a wide range of responsibilities that can vary depending on the specific company and industry. Some common responsibilities of a product manager include:

- *Defining the product vision and strategy*: The product manager is responsible for setting the long-term direction and vision for the product, and for aligning the product with the company's overall business goals and strategy. This requires a deep understanding of the customer needs and market trends, as well as the ability to prioritize and make trade-offs based on data and business impact.
- *Building and maintaining the product roadmap*: The product manager is responsible for defining and maintaining the product roadmap, which is a high-level plan that outlines the key features and milestones for the product. The roadmap serves as a guide for the development and release of the product, and it should align with the product vision and strategy. The product manager is responsible for prioritizing and scheduling the work of the cross-functional team based on the roadmap.

Gathering and analyzing market and user feedback: The product manager is responsible for gathering and analyzing market and user feedback to inform the product strategy and roadmap. This includes conducting user research and interviews, analyzing customer data and metrics, and working with the team to identify and prioritize customer needs and pain points. The product manager is also responsible for communicating this feedback to the team and stakeholders, and for incorporating it into the product development process.

- *Making data-driven decisions*: The product manager is responsible for making data-driven decisions based on customer needs, market trends, and business impact. This requires the ability to gather and analyze relevant data, to identify key metrics and performance indicators, and to make informed trade-offs and decisions based on the data.

- *Leading and motivating cross-functional teams*: The product manager is responsible for leading and motivating cross-functional teams, including engineering, design, marketing, sales, and support. This requires strong leadership and communication skills, as well as the ability to collaborate and build consensus across different functional areas. The product manager is also responsible for setting clear goals and expectations, and for providing feedback and coaching to the team to help them grow and succeed.

- *Communicating effectively with different audiences*: The product manager is responsible for communicating effectively with different audiences, including customers, stakeholders, and team members. This includes presenting and defending product plans and decisions to executives and stakeholders, as well as communicating product updates and progress to customers and the team.

In summary, a product manager is a strategic leader who is responsible for the overall direction, vision, and success of a product. They work closely with cross-functional teams and stakeholders to define the product strategy and roadmap, to gather and analyze market and user feedback, and to make data-driven decisions based on customer needs and business impact. They also lead and motivate cross-functional teams and communicate effectively with different audiences to ensure the success of the product.

WHAT IS THE DIFFERENCE BETWEEN A PRODUCT AND PROGRAM MANAGER?

Product management and program management are two critical roles in the technology industry, and they often overlap in terms of responsibilities and skills. However, there are also key differences between these roles, and it is important to understand these differences to understand which role is the best fit for you.

So, what is the difference between a product manager and a program manager?

As we discussed above, a product manager is responsible for the overall direction, vision, and success of a product. A program manager, on the other hand, is a leader who is responsible for the overall direction, coordination, and execution of a large-scale program or initiative. This may involve managing multiple projects and teams, aligning them with the overall program goals and objectives, and ensuring that they are delivered on time and within budget. The program manager is responsible for defining the program roadmap, coordinating the work of different teams and stakeholders, and tracking and reporting on program progress and performance.

There are a few key differences between product management and program management:

- *Scope*: Product management focuses on a specific product, while program management involves managing multiple projects and teams.

- *Ownership*: The product manager is the owner of the product, and they are responsible for defining the product vision and strategy. The program manager, on the other hand, does not own the individual projects and teams, but coordinates and aligns them with the overall program goals and objectives.

- *Focus*: Product management focuses on defining and delivering the product, while program management focuses on coordinating and executing the program.

- *Skills*: Product management requires a solid foundation in product development and strategy, as well as customer and market insights. Program management requires strong project management and coordination skills, as well as the ability to lead and motivate multiple teams and stakeholders.

In summary, product management and program management are two critical roles in the technology industry, and they often overlap in terms of responsibilities and skills. However, there are also key differences between these roles, including scope, ownership, focus, and skills. Understanding these differences can help you determine which role is the best fit for you.

So, which role is right for you? If you are passionate about defining and delivering high-quality products that meet the needs of customers and drive business growth, product management may be the right fit for you. If you excel at coordinating and executing large-scale programs and initiatives, and have strong project management and leadership skills, program management may be the right fit for you. The decision will depend on your strengths, interests, and career goals.

WHAT SKILLS AND QUALITIES DO SUCCESSFUL PMS HAVE?

Product management is a challenging and rewarding field that requires a diverse set of skills and qualities. To be successful as a product manager, you need to have a strong foundation in product development and strategy, as well as the ability to lead and motivate cross-functional teams and communicate effectively with different audiences.

One key skill that successful product managers have is a strong foundation in product development and strategy. This includes an understanding of the product development process, the ability to define and prioritize features, and the ability to make trade-offs based on customer needs, market trends, and business impact. Product managers also need to have a deep understanding of the customer needs and market trends, and the ability to translate these insights into a compelling product vision and roadmap.

Another important skill that product managers need is strong analytical and critical thinking skills. This includes the ability to gather and analyze data, identify key metrics and performance indicators, and make data-driven decisions based on customer needs, market trends, and business impact. Product managers also need to be able to identify and solve complex problems, and to think creatively and strategically to drive product success.

Leadership and communication skills are also critical for product managers. This includes the ability to lead and motivate cross-functional teams and communicate effectively with different audiences.

Product managers need to be able to set clear goals and expectations, provide feedback and coaching to the team, and build consensus across different functional areas. They also need to be able to present and defend product plans and decisions to executives and stakeholders, and to communicate product updates and progress to customers and the team.

Collaboration and teamwork skills are also essential for product managers. They need to be able to work effectively with cross-functional teams and stakeholders, building and maintaining strong relationships, listening, and responding to feedback, and facilitating collaboration and teamwork.

Finally, product managers need to have strong time management and organization skills, with the ability to manage their own time effectively and to coordinate the work of the cross-functional team. This includes the ability to prioritize tasks and projects based on business impact, to track and report on progress and performance, and to adjust plans and priorities as needed.

HOW CAN YOU DEMONSTRATE THESE SKILLS AND QUALITIES DURING THE INTERVIEW?

To be successful as a product manager, you need to have a strong foundation in product development and strategy, as well as the ability to lead and motivate cross-functional teams and communicate effectively with different audiences. These skills and qualities are essential for driving product success and driving business growth.

During a product management interview, it is important to demonstrate these skills and qualities to stand out from other candidates and impress the interviewer. Here are some tips for how to do this:

Prepare examples that demonstrate your foundation in product development and strategy. During the interview, the interviewer will ask you about your experience with product development and strategy. Be prepared to provide specific examples of how you have defined and prioritized features, made trade-offs based on customer needs and business impact, and translated customer needs and market trends into a compelling product vision and roadmap. This will help the interviewer see your understanding and experience in these areas.

Highlight your analytical and critical thinking skills. The interviewer will also want to see evidence of your analytical and problem-solving skills. Be prepared to discuss how you have gathered and analyzed data, identified key metrics and performance indicators, and made data-driven decisions. You can also discuss how you have identified and solved complex problems, and how you have thought creatively and strategically to drive product success. These skills are crucial for product managers, as they need to be able to make informed decisions based on data and identify and solve problems to drive product success.

Show your leadership and communication skills. Leadership and communication skills are critical for product managers, and the interviewer will be looking for evidence of these skills during the interview. Be prepared to discuss how you have led and motivated cross-functional teams, and how you have communicated effectively with different audiences. You can also talk about how you have set clear goals and expectations, provided feedback and coaching to the team, and built consensus across different functional areas. These skills are essential for product managers, as they need to be able to lead and motivate cross-functional teams and communicate effectively with different audiences to drive product success.

Discuss your collaboration and teamwork skills. Product managers also need to be strong collaborators and team players. During the interview, be prepared to discuss how you have worked effectively with cross-functional teams and stakeholders, building, and maintaining strong relationships, listening, and responding to feedback, and facilitating collaboration and teamwork. You can also talk about any specific projects or initiatives that you have worked on that required strong collaboration and teamwork skills. These skills are important for product managers, as they need to be able to work effectively with cross-functional teams and stakeholders to drive product success.

Emphasize your time management and organization skills. Finally, product managers need to have strong time management and organization skills, with the ability to manage their own time effectively and to coordinate the work of the cross-functional team. During the interview, be prepared to discuss how you have prioritized tasks and projects based on business impact, tracked, and reported on progress and performance, and adjusted plans and priorities as needed. You can also provide specific examples of how you have managed multiple projects or initiatives simultaneously, and how you have kept everything organized and on track. These skills are important for product managers, as they need to be able to manage their own time effectively and coordinate the work of the cross-functional team to drive product success.

By preparing and practicing examples that demonstrate your skills and qualities as a product manager, you can effectively showcase your strengths and impress the interviewer during your product management interview.

In summary, during a product management interview, it is important to demonstrate your foundation in product development and strategy, your analytical and problem-solving skills, your leadership and communication skills, your collaboration and teamwork skills, and your time management and organization skills. By preparing and practicing examples that showcase these skills and qualities, you can effectively showcase your strengths and impress the interviewer.

CHAPTER 3

Top PM Interview Questions

AN OVERVIEW OF THE TYPES OF QUESTIONS THAT MAY BE ASKED DURING THE PM INTERVIEW

As a product management candidate, you can expect to be asked a variety of questions during the interview process. These questions can be grouped into several categories, including behavioral questions, role-specific questions, case questions, obscure questions, and technical (code-based) questions. Here is an overview of each type of question:

Behavioral Questions: Behavioral questions are designed to assess a candidate's past experiences and how they handled specific situations. These types of questions often begin with phrases such as "Tell me about a time when..." or "Describe a situation where..." and ask you to describe a specific experience and how you approached it. Behavioral questions are used to assess your skills, abilities, and personal characteristics, and they are often used by hiring managers to better understand how you might handle similar situations in the future.

Role-specific Questions: Role-specific questions are designed to assess a candidate's fit for the specific role and responsibilities of the position. These types of questions may ask about your experience with certain types of products, your understanding of certain market segments or customer groups, or your familiarity with certain tools or technologies. Role-specific questions are used to assess your knowledge and experience in the specific areas that are relevant to the role you are applying for.

Case Questions: Case-based questions are designed to assess a candidate's problem-solving skills and ability to think on their feet. These types of questions present you with a business problem or scenario and ask you to produce a solution or recommendation. Case questions are used to assess your ability to analyze data, identify key issues, and think critically and creatively to solve problems. The case questions are sorted into five key themes:

1. Launching a new product or platform in a specific market
2. Improving the performance or success of an existing product or platform
3. Identifying and addressing issues or challenges faced by a product or platform, such as low usage, high churn, or low differentiation from competitors
4. Designing a new feature or aspect of a product or platform to meet specific goals, such as increasing user adoption, driving revenue growth, or appealing to a specific target market

5. Ensuring the success of a product or platform in a new market or demographic.

Obscure Questions: Obscure questions are designed to test a candidate's creativity, outside-the-box thinking, and ability to think on their feet. These types of questions may not have a clear or obvious answer, and they may be designed to see how you approach and solve unconventional problems. Obscure questions are used to assess your ability to think creatively and think creatively to produce innovative solutions. These are usually sorted into four major themes:

1. Imagining oneself in different scenarios or situations, such as being a character in a book or movie, having a specific superpower, or living in a different time or location
2. Choosing between different options or possibilities, such as having dinner with a historical figure, switching lives with a specific person, or traveling to a specific planet
3. Exploring personal preferences, such as which animal one would want to be, which hobby or pastime one would enjoy, or which fictional character one would want to be
4. Thinking about the motivations behind one's choices, such as why one would choose a specific character, superpower, or location.

Technical (code-based) Questions: In general, product managers do not need to be expert programmers or have a deep understanding of coding languages. However, it can be helpful for product managers to have some understanding of programming concepts and the technical capabilities and limitations of the products they are responsible for. This understanding can help product managers communicate more effectively with their engineering teams and make more informed decisions about the development of their products. Additionally, some product management roles may require more technical expertise, particularly in companies that are heavily focused on software development. In these cases, it may be necessary for product managers to have a solid foundation in programming and the ability to write code at a basic level.

Brainteaser-based Questions: These types of questions present you with a puzzle or brainteaser and ask you to solve it. Brainteaser-based questions are often used to assess a candidate's problem-solving skills, creativity, and ability to think on their feet. They may also be used to assess a candidate's ability to break down complex problems into smaller, more manageable parts and to approach problems in a logical and systematic way. They may also be used to assess a candidate's ability to analyze data, identify key issues, and think critically and strategically to solve problems.

Overall, the specific requirements for product managers will vary depending on the company and the specific role. It is important for product managers to have a clear understanding of the technical requirements of their role and to be proactive in acquiring any necessary technical skills.

THE STAR METHOD

The STAR method is a structured approach for answering behavioral interview questions, which are questions that ask you to describe specific situations you have encountered in the past and how you handled them. Behavioral interview questions are designed to assess your skills, abilities, and personal characteristics, and they are often used by hiring managers to better understand how you might handle similar situations in the future.

The STAR method is an acronym that stands for Situation, Task, Action, Result. It helps you organize your thoughts and clearly articulate your experiences, skills, and achievements in a logical and concise manner. Here is an overview of the STAR method:

- *Situation*: Begin by describing the context of the situation you were in. This could be a specific project you worked on, a challenge you faced, or a problem you needed to solve. Be sure to provide enough detail so that the interviewer understands the context and scope of the situation.
- *Task*: Next, describe the task or goal towards which you were working. What was your role or responsibility in this situation? What were you trying to achieve?
- *Action*: Now, describe the specific actions you took to achieve the task or goal. Be specific and provide concrete examples of what you did. This is the most important part of the STAR method, as it allows the interviewer to see how you approached the situation and what you did to achieve the desired result.
- *Result*: Finally, describe the result or outcome of your actions. What happened because of your efforts? Did you achieve the task or goal? Did you overcome the challenge or solve the problem? Be sure to quantify the result whenever possible, using numbers or percentages to provide a clear and objective measure of your success.

By following the STAR method, you can clearly and concisely articulate your experiences, skills, and achievements in a structured and logical manner. This can help you stand out from other candidates and effectively communicate your value to the interviewer.

101 TOP PM INTERVIEW QUESTIONS
Behavioral Questions

1. "Tell me about a time when you had to lead a team through a difficult decision-making process. What steps did you take to ensure that everyone's input was considered and what was the outcome?"

2. "Describe a situation where you had to work with a team that was not aligned on the direction of the project. How did you approach the situation and what was the outcome?"

3. "Tell me about a time when you had to solve a complex problem under a tight deadline. How did you approach the problem and what was the outcome?"

4. "Describe a situation where you had to work with a team that was not motivated. How did you approach the situation and what was the outcome?"

5. "Tell me about a time when you had to lead a team through a change in direction. How did you communicate the change to the team and what was the outcome?"

6. "Describe a situation where you had to work with a team that was experiencing conflicts. How did you approach the situation and what was the outcome?"

7. "Tell me about a time when you had to present to a group of stakeholders and make the case for your product idea. How did you prepare for the presentation and what was the outcome?"

8. "Describe a situation where you had to work with a team that was not aligned on the project goals. How did you approach the situation and what was the outcome?"
9. "Tell me about a time when you had to solve a complex problem that required creative thinking. How did you approach the problem and what was the outcome?"
10. "Describe a situation where you had to work with a diverse team. How did you ensure that everyone's voices were heard and what was the outcome?"
11. "Tell me about a time when you had to make a difficult decision that had a negative impact on the team. How did you communicate the decision to the team and what was the outcome?"
12. "Tell me about a time when you had to pivot the direction of a project mid-stream. How did you communicate the change to the team and what was the outcome?"
13. "Describe a situation where you had to work with a team that was not aligned on the project timeline. How did you approach the situation and what was the outcome?"
14. "Tell me about a time when you had to negotiate with a stakeholder or cross-functional team. How did you approach the negotiation and what was the outcome?"
15. "Describe a situation where you had to work with a team that was experiencing low morale. How did you approach the situation and what was the outcome?"
16. "Tell me about a time when you had to present to a group of executives or senior leaders. How did you prepare for the presentation and what was the outcome?"

17. "Describe a situation where you had to work with a team that was struggling to meet deadlines. How did you approach the situation and what was the outcome?"

18. "Tell me about a time when you had to manage conflicting priorities. How did you prioritize tasks and what was the outcome?"

19. "Describe a situation where you had to work with a team that was not meeting the quality standards of the project. How did you approach the situation and what was the outcome?"

20. "Tell me about a time when you had to manage up to a difficult manager or stakeholder. How did you approach the situation and what was the outcome?"

21. "Describe a situation where you had to work with a team that was not collaborating effectively. How did you approach the situation and what was the outcome?"

22. "Tell me about a time when you had to manage a budget for a project. How did you allocate resources and what was the outcome?"

23. "Describe a situation where you had to work with a team that was not meeting the project scope. How did you approach the situation and what was the outcome?"

24. "Tell me about a time when you had to manage a project that faced significant challenges or roadblocks. How did you approach the problem and what was the outcome?"

25. "Describe a situation where you had to work with a team that was not delivering on its commitments. How did you approach the situation and what was the outcome?"

Role-Specific Questions

1. "How do you approach defining and communicating the product vision and strategy to the team?"
2. "What are your strategies for building and maintaining relationships with cross-functional teams?"
3. "How do you approach gathering and analyzing customer feedback in order to inform product decisions?"
4. "What are your strategies for ensuring that everyone on the team is aligned and working towards the same goals?"
5. "How do you go about resolving conflicts within cross-functional teams?"
6. "What is your approach to communicating with customers and other external stakeholders?"
7. "How do you work with design and engineering teams to ensure that the product is delivered on time and within budget?"
8. "What is your approach to gathering and analyzing data in order to inform product decisions?"
9. "How do you go about presenting to stakeholders and making the case for your product ideas?"
10. "What are your strategies for adapting to change and leading teams through change?"
11. "How do you approach managing the product roadmap and prioritizing features?"
12. "What are your strategies for managing the budget and allocating resources for the product?"
13. "How do you go about measuring the success of a product and using data to inform future product decisions?"

14. "What are your strategies for working with cross-functional teams to ensure that the product meets the needs of the customer?"
15. "How do you approach managing the product lifecycle and ensuring that the product is successful in the market?"
16. "What are your strategies for building and maintaining relationships with customers and external stakeholders?"
17. "How do you approach managing the product launch process and ensuring a successful rollout?"
18. "What are your strategies for managing the product roadmap in a fast-paced, constantly evolving market?"
19. "How do you approach gathering and analyzing market and competitive data in order to inform product decisions?"
20. "What are your strategies for managing the product roadmap and aligning it with the overall business strategy?"
21. "What are your strategies for managing the product lifecycle and ensuring that the product remains relevant in the market?"
22. "How do you approach gathering and analyzing customer feedback in order to continuously improve the product?"
23. "What are your strategies for managing the product roadmap and adapting to changing market conditions?"
24. "How do you approach building and maintaining relationships with cross-functional teams and ensuring that the product meets the needs of all stakeholders?"
25. "How do you use data analysis and visualization tools such as Excel, Tableau, or R to inform product decisions, and can you provide an example of a specific problem you solved using these tools?"

Case Questions

1. "You are the PM for a new subscription-based meal delivery service launching in a highly competitive market. What steps would you take to differentiate your service from competitors and ensure its success?"

2. "You are the PM for an e-commerce platform that is launching in a new market. What steps would you take to ensure that the platform is successful in this new market?"

3. "You are the PM for a social media platform that has seen a decline in user engagement over the past year. What steps would you take to identify the root cause of the decline and implement a plan to increase engagement?"

4. "You are the PM for an online marketplace that is struggling to monetize. What steps would you take to identify new monetization opportunities and implement them effectively?"

5. "You are the PM for a messaging app that has seen a decline in usage over the past few months. What steps would you take to understand the cause of the decline and implement a plan to increase usage and engagement?"

6. "You are the PM for a ride-sharing app that is experiencing high churn rates among drivers. What steps would you take to understand the root causes of the churn and implement a plan to reduce it and improve driver retention?"

7. "Your company's existing video streaming service has low user adoption and usage. How would you design a new feature for the service that increases user adoption and usage?"

8. "You are the PM for a new social media platform that is launching in the next few months. You have just received feedback from a focus group indicating that the platform is not meeting

the needs of the target audience. What steps would you take to address this issue before the launch?"

9. "You are the PM for a streaming video platform that is launching in a new country. What steps would you take to understand the preferences and expectations of the target market in this country, and how would you tailor the platform to meet those needs?"

10. "You are the PM for a home security app that is launching in a new country. What steps would you take to research the market, understand the needs and preferences of the target audience, and tailor the app to meet those needs?"

Obscure Questions

1. "If you were a character in a book or movie, who would you be and why?"
2. "If you could have dinner with any historical figure, who would it be and why?"
3. "If you could be any animal, what would you be and why?"
4. "If you could have any superpower, what would it be and why?"
5. "If you could travel back in time, what period would you go to and why?"
6. "If you could switch lives with any person for a day, who would it be and why?"
7. "If you could have any job in the world, what would it be and why?"
8. "If you could live in any city in the world, where would it be and why?"
9. "If you could be any fictional character, who would you be and why?"

10. "If you could have any hobby or pastime, what would it be and why?"
11. "If you could travel to any planet, which one would you choose and why?"
12. "If you could be any age for the rest of your life, what age would you choose and why?"
13. "If you could live in any era, which one would you choose and why?"
14. "If you could have any talent or skill, what would it be and why?"
15. "If you could be any fictional character's sidekick, who would you choose and why?"

Technical (Code) based questions

1. How would you design an algorithm to find the maximum element in a list of integers?
2. How would you design an algorithm to sort a list of integers in ascending order?
3. How would you design an algorithm to find the shortest path between two nodes in a graph?
4. How would you design an algorithm to implement a stack data structure?
5. How would you design an algorithm to implement a queue data structure?
6. How would you design an algorithm to reverse a string?
7. How would you design an algorithm to find the least common multiple of two integers?

8. How would you design an algorithm to find the factorial of a given number?

9. How would you design an algorithm to check if a given string is a palindrome?

10. How would you design an algorithm to find the nth Fibonacci number?

Scenario Based Technical (Code) Questions

1. Your company has a large database of customer records. How would you design an algorithm to find the customer with the highest lifetime value?

2. Your company is launching a new product and wants to send promotional emails to a targeted list of customers. How would you design an algorithm to select the appropriate customers based on their purchase history and preferences?

3. Your company is developing a recommendation engine for a streaming service. How would you design an algorithm to recommend similar items to a user based on their past viewing history?

4. Your company is building a chatbot to assist customers with their orders. How would you design an algorithm to handle multiple concurrent conversations and route them to the appropriate customer service representative?

5. Your company is building a ride-sharing app and wants to optimize the matching of riders with drivers. How would you design an algorithm to find the best match based on location, availability, and ratings?

6. Your company is building a language translation app and wants to optimize the translation process. How would you design an

algorithm to choose the best translation for a given word or phrase based on context and past translations?

7. Your company is developing a fraud detection system for online transactions. How would you design an algorithm to identify suspicious activity and flag it for further review?

8. Your company is building a virtual personal assistant to help users manage their schedules and tasks. How would you design an algorithm to understand and interpret user requests and provide appropriate responses?

9. Your company is building a customer support platform and wants to optimize the routing of incoming inquiries to the appropriate team or agent. How would you design an algorithm to classify and route inquiries based on topic and complexity?

10. Your company is building an e-commerce platform and wants to optimize the search functionality for users. How would you design an algorithm to return the most relevant search results based on user queries and past searches?

Brainteaser Questions

1. "You have a balance scale and 10 balls. 9 of the balls weigh the same, but one of them weighs slightly more or less. How would you find the ball that is different using the scale only twice?"

2. "You are given two boxes, one containing only apples and the other containing only oranges. The boxes are labeled incorrectly, and you cannot see inside them. How can you determine which box contains the apples and which box contains the oranges using the scale only twice?"

3. "You are given a stack of cards, each with a number on one side and a letter on the other side. You are allowed to flip over two

cards at a time. How would you determine which cards have the same number or letter, using the least number of flips?"

4. "You are given two ropes and a lighter. Each rope takes exactly one hour to burn, but they do not burn at a consistent rate. One rope may burn faster or slower than the other. How would you measure out 45 minutes using only these two ropes and the lighter?"

5. "You are given a pile of marbles, some of which are red and some of which are green. You are allowed to pick up two marbles at a time. How would you determine the color of all the marbles in the least number of picks?"

6. "You are given a set of three light bulbs and three switches. Each switch controls one of the light bulbs, but you do not know which switch controls which light bulb. How would you determine the correct switch for each light bulb in the fewest number of tries?"

CHAPTER 4

Behavioral Questions Answers and Explanations

Q1: Tell me about a time when you had to lead a team through a difficult decision-making process. What steps did you take to ensure that everyone's input was considered and what was the outcome?

Example response:

One time that comes to mind was when I was leading the development of a new product feature. The team was divided on which direction we should take the feature, with some members advocating for one approach and others advocating for a different approach. As the product manager, it was my responsibility to ensure that we made the best decision for the product and the business.

To facilitate the decision-making process, I took the following steps:

1. I clearly defined the problem we were trying to solve and the goals we were trying to achieve. This helped everyone understand the context and the importance of the decision we were making.
2. I gathered data and insights from various sources, including customers, market research, and internal stakeholders. This helped us make a more informed decision based on objective evidence.
3. I facilitated a team discussion to explore different options and gather input from all team members. I made sure to encourage open and honest dialogue, and I actively listened to everyone's perspectives and ideas.
4. I synthesized all the input and evaluated the different options based on their alignment with our goals and the available data. I presented my recommendation to the team and explained my reasoning.

The outcome of this process was that we were able to make a well-informed and consensus-based decision that aligned with our goals and the needs of our customers. The team felt that their input was valued and that we had made the best decision for the product and the business.

Q2: Describe a situation where you had to work with a team that was not aligned on the direction of the project. How did you approach the situation and what was the outcome?

Example response:
One time that comes to mind was when I was working on a project to launch a new e-commerce feature for our company's website. The team was divided on which direction we should take the feature, with some members advocating for a more streamlined checkout process

and others advocating for a more personalized shopping experience. As the product manager, it was my responsibility to ensure that we were aligned on the direction of the project and that we were working towards a common goal.

To address the situation, I took the following steps:

1. I facilitated a team meeting with our lead developer, Sarah, and our UX designer, Matt, to understand the different perspectives and viewpoints on the direction of the project. I made sure to listen actively and to encourage open and honest dialogue.
2. I synthesized all the input and identified the common themes and concerns. I presented a summary of the different viewpoints to the team to ensure that everyone was on the same page.
3. I worked with the team to define a clear set of objectives and priorities for the project. I made sure that these objectives were aligned with the overall goals of the product and the business.
4. I created a project plan that outlined the specific tasks and milestones that needed to be completed to achieve our objectives. I made sure to involve the team in the planning process and to get their input and buy-in.

The outcome of this process was that we were able to align the team on the direction of the project and to work towards a common goal. The team felt that their input was valued and that we had a clear and achievable plan in place. As a result, we were able to successfully launch the new e-commerce feature on time and within budget. The feature has since been well-received by our customers and has helped to increase our sales and revenue.

Q3: **Tell me about a time when you had to solve a complex problem under a tight deadline. How did you approach the problem and what was the outcome?**

Example response:

One time that comes to mind was when I was working on a project to launch a new mobile app for our company. We had a tight deadline of two weeks to get the app ready for launch, and we encountered a complex problem with the user authentication system. The system was not functioning properly, and we were at risk of missing our launch deadline if we could not find a solution.

To address the problem, I worked with our lead developer, John, to understand the root cause of the problem and to identify potential solutions. I also gathered data and insights from our customer support team and our users to understand the impact of the problem on our customers and to identify potential workaround solutions. I then facilitated a team meeting with John and our UX designer, Rachel, to brainstorm and evaluate different options for solving the problem. After synthesizing the input from the team, I developed a plan to implement the most promising solution and worked closely with John to oversee the implementation of the solution and to ensure that it was delivered on time.

The outcome of this process was that we were able to solve the problem and to launch the mobile app on time. The app has since been well-received by our customers and has helped to increase our user engagement and retention. I learned from this experience the importance of being proactive in addressing complex problems and the value of working collaboratively with a team to find a solution.

Q4: Describe a situation where you had to work with a team that was not motivated. How did you approach the situation and what was the outcome?

Example response:

One time that comes to mind was when I was working on a project with a team of developers to launch a new feature for our company's website. The team was feeling demotivated and unenthusiastic about the project, and I was concerned that we were at risk of falling behind schedule.

To address the situation, I took the following steps:

1. I scheduled one-on-one meetings with each team member to understand their concerns and to identify any underlying issues that might be impacting their motivation.
2. I synthesized the input from the team and identified common themes, such as a lack of clear direction and a lack of resources.
3. I worked with my manager to address these issues and to provide the team with the support and resources they needed to be successful.
4. I facilitated team meetings to discuss the progress of the project and to ensure that everyone was aligned on the goals and priorities.
5. I recognized and celebrated the team's achievements along the way, including sending out a company-wide email to recognize their arduous work.

The outcome of this process was that the team was able to get back on track and to successfully launch the new feature on time. The team reported feeling more motivated and engaged with the project, and we received positive feedback from our users. I learned from this experience the importance of listening to and supporting team members and the value of recognizing and celebrating their achievements.

Q5: **Tell me about a time when you had to lead a team through a change in direction. How did you communicate the change to the team and what was the outcome?**

Example response:

One time that comes to mind was when I was leading a team of product managers on a project to develop a new software platform. We had been working on the project for several months and had made considerable progress, but halfway through the project, we received added information that required us to change the direction of the project.

To address the change in direction, I first gathered and analyzed the added information and determined the implications for the project. I then met with my manager, Jane, to discuss the situation and to get her input and support. After that, I called a team meeting and explained the situation to the team, including the added information we had received and the implications for the project. I led a discussion with the team to gather their input and ideas on how to approach the change in direction, and then developed a revised plan for the project based on the input from the team and my manager. Finally, I communicated the revised plan to the team.

The outcome of this process was that the team was able to successfully adapt to the change in direction and to continue making progress on the project. We were able to deliver the software platform on time and within budget, and the product was well-received by our customers. I learned from this experience the importance of being adaptable and flexible and the value of communicating and collaborating with a team to navigate change.

Q6: Describe a situation where you had to work with a team that was experiencing conflicts. How did you approach the situation and what was the outcome?

Example response:
One situation that comes to mind was when I was working on a project with a team of designers and developers to create a new mobile app. The team had been working together for several months, but we started to experience conflicts and tension as we neared the end of the project.

To address the conflicts, I took the following steps:

1. I identified the root causes of the conflicts, which included misunderstandings about roles and responsibilities, differing viewpoints on design and functionality, and time constraints.
2. I met with my manager, Tom, to discuss the situation and to get his input and support.
3. I called a team meeting and led a discussion with the team to understand the different perspectives and to identify common goals and priorities.
4. I facilitated a mediation session with the team to resolve the conflicts and to align on a plan for completing the project.

5. I worked with the team to develop a revised project plan that incorporated the input from the mediation session and that included clear roles and responsibilities, a defined scope and timeline, and a process for managing conflicts.
6. The outcome of this process was that the team was able to resolve the conflicts and to successfully complete the project on time and within budget. The mobile app was well-received by our customers and received positive reviews. I learned from this experience the importance of identifying and addressing conflicts early and the value of facilitating mediation sessions to resolve conflicts and to align on a common goal.

Q7: **Tell me about a time when you had to present to a group of stakeholders and make the case for your product idea. How did you prepare for the presentation and what was the outcome?**

Example response:

One situation that comes to mind was when I had to present to the executive team at XYZ Company and make the case for a new product idea that I had developed. The product was a software platform that would help small businesses manage their finances more efficiently.

To prepare for the presentation, I took the following steps:

1. I researched the market and gathered data on the needs and pain points of small businesses regarding financial management.
2. I worked with the engineering team, led by Sarah, to develop a prototype of the software platform and to gather feedback from potential customers.

3. I created a PowerPoint presentation that included an overview of the market research, a demonstration of the prototype, and a financial analysis showing the potential return on investment for the product.
4. I practiced the presentation with Sarah and with a few colleagues to get feedback and to fine-tune the content and delivery.
5. I reached out to a few key stakeholders ahead of the presentation to provide them with a preview of the product and to solicit their feedback and support.

The outcome of the presentation was that the executive team was very impressed with the product idea and the market research and data that I presented. They gave me the green light to proceed with developing the product and to bring it to market. The product was ultimately launched and was very successful, generating significant revenue for the company. I learned from this experience the importance of thoroughly researching and preparing for a presentation, as well as the value of seeking feedback and support from key stakeholders ahead of time.

Q8: Describe a situation where you had to work with a team that was not aligned on the project goals. How did you approach the situation and what was the outcome?

Example response:
One situation that comes to mind was when I was working on a product launch at ABC Company. I was part of a cross-functional team that included marketing, sales, and customer support, and our goal was to launch a new line of fitness products in time for the holiday season.

However, as we got closer to the launch date, I noticed that there were some conflicting ideas and priorities among the team members.

To address this situation, I took the following steps:

1. I called a team meeting to assess the current state of the project and to identify any bottlenecks or roadblocks.
2. I facilitated a discussion among the team members to gather their feedback and ideas about the launch.
3. I listened to everyone's perspectives and tried to understand their concerns and priorities.
4. I synthesized the team's feedback and proposed a revised plan that incorporated everyone's ideas and addressed their concerns.
5. I communicated the revised plan to the team and sought their buy-in and commitment to the project.

The outcome of this process was that the team was able to align on the project goals and to work together more effectively. We were able to successfully launch the new line of fitness products on time and to exceed our sales targets. I learned from this experience the importance of regularly checking in with the team and soliciting their feedback, and the value of being flexible and adaptable in the face of changing circumstances.

Q9: Tell me about a time when you had to solve a complex problem that required creative thinking. How did you approach the problem and what was the outcome?

Example response:

One situation that comes to mind was when I was working on a project at DEF Company to develop a new social media platform for young adults. The project team was facing a number of challenges, including a tight deadline, limited resources, and a highly competitive market.

One of the biggest challenges we faced was how to differentiate our platform from existing competitors and to create a unique value proposition for our target audience. We needed to produce a creative solution that would set us apart and make our platform stand out.

To approach this problem, I took the following steps:

1. I gathered a diverse group of team members from different departments and asked them to brainstorm ideas for the platform. I made sure to include team members from marketing, design, and engineering, as well as representatives from our target audience.

2. I encouraged everyone to think creatively and to challenge assumptions about what our target audience wanted. I used techniques like lateral thinking and the "five whys" to help stimulate creative thinking and to identify new and unconventional ideas.

3. I facilitated a discussion among the team members to gather their feedback and ideas. I made sure to listen actively and to provide constructive feedback on their ideas, and I encouraged everyone to share their thoughts and perspectives.

4. I synthesized the team's feedback and proposed a new concept for the platform that incorporated a number of creative and innovative features. These features included a personalized recommendation system, a social networking component, and a gamification element that rewarded users for engaging with the platform.

5. I presented the concept to the project sponsor and the senior leadership team and made the case for why it was a strong and compelling solution to the problem. I used data and customer insights to support my argument, and I highlighted the potential benefits of the concept for the company and for our target audience.

The outcome of this process was that the concept was approved, and we were able to move forward with the project. The platform was successfully launched and received a great deal of positive feedback from users. I learned from this experience the value of gathering diverse perspectives and encouraging creative thinking, and the importance of being able to communicate and sell a creative solution to stakeholders. I also learned the value of using techniques like lateral thinking and the "five whys" to stimulate creative thinking and to identify unconventional solutions to complex problems.

Q10: Describe a situation where you had to work with a diverse team. How did you ensure that everyone's voices were heard and what was the outcome?

Example response:
I had the opportunity to work on a diverse team at XYZ Company, where we were tasked with launching a new product in a highly

competitive market. One of the challenges we faced was ensuring that everyone's voices were heard and that we were able to effectively collaborate as a team despite our diverse backgrounds and perspectives.

To address this challenge, I implemented a number of strategies. First, I made a conscious effort to actively listen to each team member and actively solicit their feedback and ideas. I also set up regular team meetings where everyone had the opportunity to contribute their thoughts and ideas, and I encouraged the team to engage in open and honest communication.

Additionally, I implemented a rotation system where each team member had the opportunity to lead the team in a specific task or project. This allowed everyone to have a voice in the decision-making process and ensured that everyone had the opportunity to contribute their unique skills and perspectives.

These efforts paid off and we were able to successfully launch the product on time and within budget. The diverse perspectives and skills of the team members were integral to the success of the project and allowed us to approach problems from multiple angles and produce creative solutions.

Q11: **Tell me about a time when you had to make a difficult decision that had a negative impact on the team. How did you communicate the decision to the team and what was the outcome?**

Example response:
One of the most difficult decisions I had to make as a product manager at ABC Company was when we had to pivot our product strategy due to changing market conditions. While this decision was necessary for

the long-term success of the product, it also had a negative impact on the team as it meant that we had to let go of a number of features that the team had worked hard on and were emotionally invested in.

To communicate this decision to the team, I held a meeting where I explained the reasoning behind the decision and the data that supported it. I also made sure to acknowledge the hard work and effort that the team had put in and offered my support as we transitioned to the new strategy.

While the initial reaction to the decision was understandably negative, I worked closely with the team to help them understand the long-term benefits of the change and how it would ultimately help us achieve our goals. I also made sure to keep the team informed of our progress and provided regular updates on our progress.

In the end, the team was able to successfully pivot, and the product went on to achieve remarkable success in the market. While it was a difficult decision, it was the right one for the long-term success of the product and the team.

Q12: Tell me about a time when you had to pivot the direction of a project mid-stream. How did you communicate the change to the team and what was the outcome?

Example response:
One time, I was working on a project with a team at XYZ Company. Our project was to develop a new software application, but halfway through the process, we realized that the technology we were using was not going to be sufficient for the needs of our clients.

I knew we had to pivot the direction of the project, so I gathered the team together and explained the situation. I outlined the new direction we needed to take and the reasons for the change. I also emphasized that this was an opportunity to create an even better product for our clients.

To communicate the change effectively, I made sure to listen to the concerns and ideas of the team and involve them in the decision-making process. We worked together to produce a revised plan and timeline for the project.

The outcome was a success. The revised software application received positive feedback from our clients and exceeded their expectations. The team also felt more invested in the project and was able to work together effectively to overcome the challenge.

Q13: Describe a situation where you had to work with a team that was not aligned on the project timeline. How did you approach the situation and what was the outcome?

Example response:
While working at GHI Company, I had to work with a team that was not aligned on the project timeline for a major product launch. Some team members were concerned about meeting the tight deadline, while others felt that we had sufficient time and resources to complete the project on time.

To address this challenge, I held a team meeting where we discussed the timeline and the specific tasks and milestones that needed to be completed to meet the deadline. I also made sure to clearly outline the consequences of not meeting the deadline and the potential impact on the business.

To help ensure that we were able to meet the deadline, I implemented a more structured project management approach and set up regular check-ins with the team to track progress and identify any potential roadblocks. I also made sure to provide the team with the necessary resources and support they needed to complete their tasks on time.

In the end, we were able to successfully launch the product on time and within budget, and the team was able to come together and work effectively towards our common goal. This experience taught me the importance of clear communication and effective project management in ensuring that a team is aligned and able to successfully meet deadlines.

Q14: Tell me about a time when you had to negotiate with a stakeholder or cross-functional team. How did you approach the negotiation and what was the outcome?

Example response:
I had to negotiate with a cross-functional team at JKL Company when we were working on a new product launch and there was disagreement about the budget and resources allocated to the project. Some team members felt that we needed more resources to deliver on the project goals, while others felt that we had sufficient resources and that any additional investments would not be justified.

To approach the negotiation, I first gathered data and research to support my position and to clearly outline the potential benefits and risks of different resource allocation scenarios. I also made sure to listen to the concerns and perspectives of all team members and to consider their ideas and suggestions.

I then held a meeting with the team where I presented my findings and recommendations, and we were able to have an open and honest discussion about the best course of action. In the end, we were able to reach a compromise and the team agreed to allocate additional resources to the project, with the understanding that we would need to track progress closely and adjust as needed.

This experience taught me the importance of being well-prepared and having a clear understanding of the needs and priorities of all stakeholders to effectively negotiate and reach a mutually beneficial outcome. The product launch was a success, and the team was able to deliver on the project goals within the agreed-upon timeline and budget.

Q15: Describe a situation where you had to work with a team that was experiencing low morale. How did you approach the situation and what was the outcome?

Example response:
I had to work with a team at XYZ Company that was experiencing low morale when we were working on a high-stress project with tight deadlines. Some team members were feeling overwhelmed and burned out, and there was a general sense of frustration and negative energy within the team.

To address this challenge, I first held one-on-one conversations with each team member to understand their specific concerns and to identify any potential issues that needed to be addressed. I also made sure to provide the team with clear communication and support, and to recognize and appreciate their efforts and contributions.

To help boost morale, I also implemented a number of team-building activities and initiatives, such as team lunches and team bonding activities, to help foster a more positive and collaborative work environment. I also made sure to set realistic goals and expectations for the team and to provide them with the resources and support they needed to succeed.

In the end, the team was able to come together and work effectively towards our common goal, and we were able to successfully complete the project on time and within budget. This experience taught me the importance of effective communication and team building in creating a positive and productive work environment.

Q16: Tell me about a time when you had to present to a group of executives or senior leaders. How did you prepare for the presentation and what was the outcome?

Example response:
One time, I was asked to present to a group of executives at XYZ Company about a technical project I had been leading. The project was to implement a new database system, and I knew that the executives would be expecting a detailed and technical presentation.

To prepare for the presentation, I took the following steps:

1. I gathered all the relevant technical information and data about the project, including details about the database system, the

benefits of the implementation, and any potential challenges we had encountered.

2. I created a clear and concise outline for the presentation, highlighting the key technical aspects of the project and the results we had achieved.

3. I practiced my delivery and timed the presentation to ensure it was within the allotted period.

4. I created visual aids, such as technical diagrams and charts, to help illustrate the technical details of the project and make the presentation more engaging.

The presentation went well, and the executives were impressed with the level of technical knowledge and detail I was able to provide. The outcome was a success, and the new database system was successfully implemented. The executives also gave me positive feedback on my technical presentation skills and asked me to present at future company events.

Q17: **Describe a situation where you had to work with a team that was struggling to meet deadlines. How did you approach the situation and what was the outcome?**

Example response:
I had to work with a team at DEF Company that was struggling to meet deadlines on a software development project. We were under a tight deadline and the team was facing a number of challenges, including resource constraints and technical issues.

To address this challenge, I first conducted a thorough review of the project plan and identified any potential bottlenecks or inefficiencies that were causing delays. I then worked with the team to develop a revised plan that focused on prioritizing the most critical tasks and allocating resources more effectively.

I also implemented a number of process improvements, such as introducing agile methodologies and implementing more effective project management tools and systems, to help streamline the team's workflow and increase productivity. I also made sure to provide the team with the support and resources they needed to overcome any technical challenges they were facing.

In the end, the team was able to successfully meet the deadlines and we were able to deliver the software on time. This experience taught me the importance of effective project management and the value of process improvements in helping teams to overcome challenges and meet deadlines.

Q18: Tell me about a time when you had to manage conflicting priorities. How did you prioritize tasks and what was the outcome?

Example response:
At XYZ Company, I had to manage conflicting priorities on a software development project that was under a tight deadline. The project required us to deliver a number of features and functionality within a short timeframe, and we were facing competing demands from different stakeholders.

To address this challenge, I conducted a thorough review of the project plan and identified the most critical tasks that needed to be completed first. I used a combination of tools and techniques, such as a Kanban board and impact mapping, to prioritize these tasks based on their importance and impact on the project.

I also engaged with the various stakeholders to understand their priorities and helped to align the team's work with those priorities. This required me to be proactive in communicating with the stakeholders and ensuring that they were kept informed of our progress and any challenges we were facing. I used a variety of communication channels, such as regular status updates and one-on-one meetings, to keep everyone on the same page.

In the end, we were able to successfully deliver the project on time and meet the expectations of all stakeholders. This experience taught me the importance of effective task prioritization and stakeholder management in helping to balance competing priorities and deliver successful projects. It also demonstrated my ability to adapt and utilize different tools and techniques to effectively manage complex projects.

Q19: Describe a situation where you had to work with a team that was not meeting the quality standards of the project. How did you approach the situation and what was the outcome?

Example response:
As a technical product manager at ABC Company, I worked on a project to develop a new mobile app. During the development process, I noticed that the team was struggling to meet the quality standards we had set for the project. There were a number of bugs and issues that were causing delays and hindering the user experience.

I knew that inferior quality could have a negative impact on the project, so I approached the situation by identifying the root causes of the issue. I gathered the development team together and asked them to share their thoughts and feelings about the project and the challenges they were facing. I listened to their concerns and ideas and tried to understand the underlying technical issues that were causing the quality issues.

To address the issue, we implemented a number of solutions. We clarified the quality standards and expectations for the project and provided additional resources, such as additional testing and debugging resources. We also held regular team meetings to review the progress of the project and address any issues that arose.

The outcome was a success. The development team was able to improve the quality of the mobile app and launch it on time. The team also felt more organized and cohesive, and we were able to work together effectively to overcome the technical challenges we faced.

One key lesson I learned from this experience was the importance of setting clear quality standards and expectations from the beginning of a project. I also learned the value of regularly reviewing progress and addressing issues as they arise, as well as the importance of providing the necessary resources and support to the team to meet the project's goals. These lessons have helped me to be more effective as a technical product manager and to deliver successful projects in the future.

Q20: Tell me about a time when you had to manage up to a difficult manager or stakeholder. How did you approach the situation and what was the outcome?

Example response:

As a technical product manager at ABC Company, I had to work with a stakeholder who was very technical and had high expectations for the project we were working on. This stakeholder was a subject matter expert in their field and had a deep understanding of the technical details of the project.

I knew that managing up to this stakeholder would require a high level of technical expertise and a proactive approach, so I took the following steps:

1. I made sure to stay up to date on the latest technical developments and best practices in the field, and I always tried to anticipate the stakeholder's technical needs and concerns.

2. I communicated regularly with the stakeholder, providing updates on the progress of the project, and seeking feedback and guidance as needed. I made sure to present the technical details of the project in a clear and concise manner, and I was prepared to answer any technical questions or objections they had.

3. I tried to be proactive in addressing any technical issues or concerns that arose, and I always tried to present technical solutions rather than just problems.

The outcome was a success. Despite the challenges, I was able to work effectively with the stakeholder and deliver a high-quality technical product. I also learned the value of staying up to date on technical developments and being proactive in addressing technical issues in managing up to difficult technical stakeholders. These lessons have helped me to be more effective as a technical product manager and to build strong working relationships with technical stakeholders and subject matter experts.

Q21: **Describe a situation where you had to work with a team that was not collaborating effectively. How did you approach the situation and what was the outcome?**

Example response:

At XYZ Company, I was working on a software development project where the team was struggling with poor collaboration and communication. This was causing delays and leading to a lack of progress on the project.

To address this issue, I held a series of team meetings to identify the root causes of the poor collaboration and produce a plan to improve the situation. We identified several factors that were contributing to the problems, such as conflicting schedules, lack of clear project goals and objectives, and inadequate tools and resources.

To address these issues, I implemented a number of changes to our processes and practices. For example, I introduced a more flexible scheduling system that allowed team members to work around their other commitments, and I worked with the team to develop clear project goals and objectives that everyone could work towards. I also invested in new tools and resources that would enable the team to collaborate more effectively, such as a project management platform and team communication software.

As a result of these changes, we were able to significantly improve the team's collaboration and communication, and we were able to make noteworthy progress on the project. This experience taught me the importance of identifying and addressing root causes to improve team collaboration and the value of investing in the right tools and resources to support the team's efforts. It also demonstrated my ability to lead and motivate a team to achieve a common goal, particularly in the context of a software development project.

Q22: Tell me about a time when you had to manage a budget for a project. How did you allocate resources and what was the outcome?

Example response:

As a technical product manager at ABC Company, I had to manage the budget for a project to develop a new software application. The project had a limited budget and a tight timeline, so it was important to allocate resources effectively to deliver the project on time and within budget.

To manage the budget, I took the following steps:

1. I created a detailed project plan that included a budget breakdown for each task and deliverable.
2. I worked with the development team to identify the resources needed to complete each task and deliverable, and I allocated resources based on priority and importance.
3. I monitored the budget throughout the project and adjusted as needed to stay within budget and on track.

The outcome was a success. The development team was able to deliver the software application on time and within budget. The project was also well-received by the client and received positive feedback.

One key lesson I learned from this experience was the importance of careful planning and budget management in a technical project. I also learned the value of working closely with the development team to identify and allocate resources effectively, and the importance of monitoring and adjusting the budget as needed. These lessons have helped me to be more effective as a technical product manager and to deliver successful projects within budget in the future.

Q23: Describe a situation where you had to work with a team that was not meeting the project scope. How did you approach the situation and what was the outcome?

Example response:
As a technical product manager at XYZ Company, I worked on a project with a team that was struggling to meet the project scope. The team was working on the development of a new software application, and we had a number of specific requirements and deadlines that we needed to meet. However, as we progressed through the project, I noticed that the team was struggling to stay on track and meet the project scope.

I knew that not meeting the project scope could have a negative impact on the project, so I approached the situation by identifying the root causes of the issue. I gathered the development team together and asked them to share their thoughts and feelings about the project and the challenges they were facing. I listened to their concerns and ideas and tried to understand the underlying technical issues that were causing the scope issues.

To address the issue, we implemented a number of solutions. We clarified the project scope and requirements and provided additional resources and support as needed. We also held regular team meetings to review the progress of the project and address any issues that arose.

The outcome was a success. The development team was able to meet the project scope and deliver the software application on time. The team also felt more organized and cohesive, and we were able to work together effectively to overcome the technical challenges we faced.

One key lesson I learned from this experience was the importance of clearly defining the project scope and requirements from the beginning of a project. I also learned the value of regular communication and review in staying on track and meeting the project scope. These lessons have helped me to be more effective as a technical product manager and to deliver successful projects within scope in the future.

Q24: Tell me about a time when you had to manage a project that faced significant challenges or roadblocks. How did you approach the problem and what was the outcome?

Example response:
I worked on a project with a team that faced significant challenges and roadblocks. The project was a software development project that required the integration of multiple complex technical systems.

One of the main challenges we faced was a lack of clear communication and coordination among the team members. This led to delays and misunderstandings, which made it difficult to stay on track and meet the project deadlines. Additionally, we encountered several technical issues that required careful planning and problem-solving to overcome.

To address these problems, I approached the situation by identifying the root causes of the issues and implementing solutions. I gathered the development team together and asked them to share their thoughts and feelings about the project and the challenges they were facing. I listened to their concerns and ideas and tried to understand the underlying technical issues that were causing the lack of communication and coordination.

To address the issues, we implemented a number of solutions. We clarified roles and responsibilities to ensure that everyone knew exactly what was expected of them and what technical tasks they were responsible for. We also established regular team meetings to review the progress of the project and address any technical issues that arose. We also encouraged team members to collaborate and share their technical expertise and ideas, and we provided additional resources and support as needed.

The outcome was a success. Despite the challenges, we were able to deliver the project on time and within budget. The project was also well-received by the client and received positive feedback.

One key lesson I learned from this experience was the importance of clear communication and coordination in a team environment, especially in a complex and resource-intensive project. I also learned the value of encouraging collaboration and sharing of technical expertise among team members, and the importance of providing the necessary resources and support to the team to overcome technical challenges and achieve project goals. These lessons have helped me to be more effective as a technical product manager and to deliver successful projects despite facing significant challenges or roadblocks in the future.

Q25: **Describe a situation where you had to work with a team that was not delivering on its commitments. How did you approach the situation and what was the outcome?**

Example response:

One example of a situation where I had to work with a team that was not delivering on its commitments was when I was leading the development of a new mobile app for a healthcare company. The development team consisted of a mix of in-house and outsourced engineers, and we were under a tight deadline to launch the app in time for a major industry conference.

As we approached the halfway point of the project, it became clear that the team was falling behind schedule and we were at risk of missing the deadline. Upon further investigation, I discovered that there were several issues causing the delays. Some team members were not fully committed to the project and were not putting in the necessary effort, while others were struggling with the technical challenges of the project.

To address these issues, I took a number of steps. First, I met with the team to discuss the problems we were facing and to reaffirm the importance of meeting the deadline. I also made it clear that everyone was expected to put in their best effort and that any issues or concerns they had should be raised as soon as possible.

Next, I worked with the team to reorganize the project plan and reassign tasks to ensure that we were utilizing the skills and expertise of each team member to the best of their abilities. I also brought in additional resources, including an experienced project manager, to help with the technical challenges and to ensure that the team was staying on track.

In the end, these efforts paid off and we were able to launch the app on time. The conference was an enormous success, and the app received a lot of positive feedback from attendees. This experience taught me the importance of proactive problem-solving and the value of clear communication and collaboration within a team.

CHAPTER 5

Role-Specific Questions Answers and Explanations

Q1: **How do you approach defining and communicating the product vision and strategy to the team?**

Example response:

In my previous role as a Technical Product Manager at XYZ Company, I was responsible for defining and communicating the product vision and strategy to a cross-functional team of engineers, designers, and researchers.

To gather input and ensure buy-in from the team, I held regular check-ins with each member to get their perspective on the direction of the product and their ideas for how we could move forward. I also solicited feedback from key stakeholders, including customers and industry experts, through surveys, focus groups, and one-on-one interviews.

To communicate the vision and strategy, I used a variety of methods. First, I created a high-level roadmap that outlined the key milestones and deliverables for the next 6-12 months. This helped everyone understand the overall direction of the product and how their work fit into the bigger picture. I also developed detailed product specifications that outlined the specific features and functionality we needed to build, along with the reasoning behind each decision. These documents served as a reference for the team and helped everyone stay focused on the goals.

In addition to written materials, I used visual aids like diagrams, charts, and mockups to help bring the product vision to life. This made it easier for the team to understand the product and how it fit into the market.

To ensure that the team understood and was aligned with the vision and strategy, I held regular team meetings and presented updates on the progress we were making. I also made myself available to answer any questions or address any concerns they had.

Overall, effective communication was key to ensuring that everyone was on the same page and working towards the same goals. By gathering input, using a variety of tools and techniques to communicate the vision and strategy, and regularly checking in with the team, we were able to move the product forward and achieve our objectives.

Q2: **What are your strategies for building and maintaining relationships with cross-functional teams?**

Example response:
To effectively build and maintain relationships with cross-functional teams as a technical product manager, it is important to be proactive and proactive in communication.

From a soft-skills perspective, one strategy I have found to be effective is regularly scheduling check-ins with team members and stakeholders to ensure that everyone is on track and has the support they need to succeed. These check-ins can be in the form of one-on-one meetings, team meetings, or even short touch base calls. By regularly checking in, I can stay informed about the progress of projects and identify any potential roadblocks or issues early on, allowing me to address them before they become larger problems.

Another strategy I have found useful is being transparent and open about my intentions and goals as a product manager. By clearly communicating my vision and strategy for a project, I can get buy-in and support from cross-functional teams, which helps to foster a positive and collaborative working environment.

There are several additional strategies from a technical perspective that I might use to build and maintain relationships with cross-functional teams:

1. *Leveraging data and analytics:* By using data and analytics to inform decision-making, a product manager can demonstrate their technical expertise and build credibility with the team.

2. *Collaborating on code reviews:* By actively participating in code reviews, a product manager can demonstrate their technical knowledge and build strong relationships with the engineering team.
3. *Providing technical mentorship:* By providing guidance and support to team members on technical challenges, a product manager can build trust and respect within the team.
4. *Participating in technical training and development:* By taking the time to stay up to date on the latest technical trends and developments, a product manager can demonstrate their commitment to the team and build relationships through shared learning.
5. *Proactively managing technical debt:* By proactively identifying and addressing technical debt, a product manager can demonstrate their technical savvy and build trust with the engineering team.

Q3: **How do you approach gathering and analyzing customer feedback to inform product decisions?**

Example response:

As a technical product manager, I understand the importance of gathering and analyzing customer feedback to inform product decisions. One strategy I have used in the past is to set up regular customer interviews or focus groups, where I can speak directly with users and gather their feedback on the product. I use a variety of tools to analyze this feedback, including customer survey software, data visualization tools, and text analysis tools.

Another approach I have used is to set up a customer advisory board, which is a group of customers that I meet with on a regular basis to get their input and feedback on the product. This is a great way to get a diverse set of perspectives and to ensure that the product is meeting the needs of a wide range of users.

In addition to these more formal approaches, I also make sure to stay attuned to customer feedback that comes in through more informal channels, such as social media, customer service inquiries, and online reviews. I use a variety of tools and techniques to track this feedback, including sentiment analysis tools, customer service software, and social media monitoring tools.

Overall, my approach to gathering and analyzing customer feedback is to be proactive and to use a variety of tools and techniques to ensure that I am getting a complete and accurate picture of how users are interacting with the product. This allows me to make informed decisions about how to improve the product and better meet the needs of my customers.

Q4: What are your strategies for ensuring that everyone on the team is aligned and working towards the same goals?

Example response:
As a technical product manager, one of my key strategies for ensuring alignment among team members is to clearly communicate the goals and objectives of the project from the very beginning. This includes setting specific, measurable, attainable, relevant, and time-bound (SMART) goals, and regularly reviewing progress towards these goals with the team.

I also make sure to involve team members throughout the following:

1. *Clearly defining the project goals and objectives:* It is important to have a clear understanding of the project goals and objectives and to communicate them effectively to the team. This helps to ensure that everyone is working towards the same objectives and that their efforts are aligned with the overall project goals.

2. *Establishing clear roles and responsibilities:* Assigning specific roles and responsibilities to team members helps to ensure that everyone knows exactly what is expected of them and how they fit into the overall project. This helps to reduce confusion and ensure that everyone is working towards the same goals.

3. *Communicating regularly:* Regular communication is key to ensuring that everyone on the team is aligned and working towards the same goals. This can be accomplished through team meetings, one-on-one conversations, and other forms of communication, such as email updates or project management software.

4. *Encouraging collaboration and teamwork:* Encouraging team members to collaborate and share their ideas and expertise can help to ensure that everyone is working towards the same goals and that everyone's skills and knowledge are being utilized effectively.

5. *Monitoring progress and provide feedback:* Regularly monitoring the progress of the project and providing feedback to team members can help to ensure that everyone is aligned and working towards the same goals. This can be done through progress reports, team meetings, and other means of communication.

Q5: **How do you go about resolving conflicts within cross-functional teams?**

Example response:

When I was working on a project at XYZ Company, we ran into a situation where there was a conflict between the engineering team and the design team. The engineering team felt that the designs provided by the design team were not realistic given the technical constraints and timeline of the project, while the design team felt that their designs were necessary to achieve the desired user experience.

To resolve this conflict, I first made sure to listen to both sides and understand their perspectives and concerns. I then brought the team together for a meeting and facilitated a discussion where we openly and honestly talked about the issues at hand. I encouraged both sides to be open to compromise and to work towards finding a solution that would meet the needs of both teams.

In the end, we were able to produce a solution that incorporated some of the designs from the design team and considered the technical constraints and timeline of the project. This solution required some additional work and effort from both teams, but it resulted in a product that was successful in meeting both the technical requirements and the desired user experience.

Overall, my approach to resolving conflicts within cross-functional teams is to first listen to all sides and understand their perspectives, and then facilitate an open and honest discussion where everyone can work towards finding a mutually beneficial solution.

Q6: What is your approach to communicating with customers and other external stakeholders?

Example response:

My approach to communicating with customers and other external stakeholders is based on the belief that clear communication is key to building trust and maintaining strong relationships. It is important to be responsive and available to answer questions and address concerns, and to communicate updates and progress on projects in a timely and transparent manner.

One specific example of how I have effectively communicated with external stakeholders was during a project I worked on at XYZ Company. I was the technical product manager for a software development project that had a number of high-profile customers. To ensure that the customers were kept informed and involved in the project, I implemented a number of communication strategies. I held regular conference calls and webinars to update the customers on the progress of the project, and I made myself available to answer any questions or concerns they had. I also provided regular written updates and communicated through email and other online tools as needed.

The outcome of this approach was a success. The customers felt informed and engaged in the project, and they were satisfied with the level of communication they received. This helped to build trust and maintain strong relationships with the customers and contributed to the success of the project.

Overall, my approach to communicating with customers and other external stakeholders is centered on building trust and maintaining strong relationships through clear, effective, and timely communication.

Q7: How do you work with design and engineering teams to ensure that the product is delivered on time and within budget?

Example response:

There are several approaches I could take to work with design and engineering teams to ensure that the product is delivered on time and within budget. Some strategies include:

1. *Defining technical requirements:* It is important to define the technical requirements for the product early in the development process, as this helps guide the design and engineering efforts. This could include identifying the technologies and platforms that will be used, establishing performance and scalability targets, and defining any security or compliance requirements.

2. *Establishing a development workflow:* Establishing a development workflow can help ensure that the project stays on track and that code is being developed and tested in an organized and efficient manner. This could involve using a version control system, establishing a code review process, and setting up automated testing and deployment pipelines.

3. *Managing technical dependencies:* Depending on the product, there may be technical dependencies that need to be managed, such as third-party libraries or APIs. It is important to keep track of these dependencies and ensure that they are properly integrated into the product and kept up to date.

4. *Identifying and addressing technical risks:* There may be technical risks that could impact the delivery of the product, such as the complexity of the codebase or the need to integrate with unfamiliar technologies. It is important to identify and address these risks early on to avoid delays or budget overruns.

5. *Collaborating with cross-functional teams:* As a PM, it is important to work closely with the design and engineering teams, as well as other cross-functional teams such as product marketing and customer support. This can help ensure that the product meets the needs of the target audience and is delivered in a timely manner.

Finally, it is important to be flexible and adapt to changes as they arise. This could involve revising the project plan as needed or finding creative solutions to unexpected challenges.

Q8: What is your approach to gathering and analyzing data to inform product decisions?

Example response:

I was the PM for a mobile app that helped users track their daily water intake. I received customer feedback indicating that some users were having difficulty remembering to log their water intake throughout the day, and as a result, they were not meeting their daily hydration goals.

To address this issue, I decided to gather data to inform a product decision on how to improve the app's reminder feature. Here is how I approached this:

1. *Identified the data that was relevant to the product:* In this case, I wanted to gather data on how often users were forgetting to log their water intake, as well as data on which users were most likely to forget. I also wanted to gather data on the types of reminders that users found most helpful.

2. *Gathered the data:* To gather this data, I sent out a survey to a sample of users asking about their experience with the app's

reminder feature. I also reviewed analytics data on usage patterns and user demographics to identify trends or patterns.

3. *Analyzed the data:* Once I had collected the data, I analyzed it to extract insights and inform my product decision. I used Excel to create graphs and charts visualizing the data and ran statistical tests to identify trends or patterns.

4. *Communicated the results:* After analyzing the data, I communicated the results to the relevant teams and stakeholders. This involved presenting the findings in a clear and concise manner, highlighting key takeaways and recommendations, and discussing any implications for the product.

5. *Iterated and refined:* Finally, I continued to gather and analyze data on an ongoing basis to refine the product decision and ensure that the app's reminder feature was meeting the needs of users. This involved setting up regular data-gathering processes, such as customer surveys or analytics tracking, and using that data to inform ongoing product development efforts.

Finally, it is important to be flexible and adapt to changes as they arise. This could involve revising the project plan as needed or finding creative solutions to unexpected challenges.

Q9: How do you go about presenting to stakeholders and making the case for your product ideas?

Example response:
One specific experience I had was when I was leading the development of a new product for XYZ Company. We were seeking funding from the executive team to move forward with the project, so I knew

I needed to make a compelling case for why this product was worth investing in.

First, I made sure to thoroughly research the market and understand the needs of our target customers. I gathered data on related products in the industry, as well as customer feedback from focus groups and surveys. This helped me to identify the unique value proposition of our product and how it would solve a pain point for our customers.

Next, I worked with my team to create a comprehensive presentation that highlighted the key features and benefits of the product, as well as the potential return on investment. We also included a detailed project plan and timeline, to show that we had a clear vision for how to execute on the project.

During the presentation, I walked the stakeholders through the live demo and highlighted the key features and benefits of the product. I also addressed any questions or concerns they had and provided additional supporting data as needed.

Our presentation was successful, and we were able to secure the funding we needed to move forward with the project. The product ended up being a huge success and generated significant revenue for the company.

Q10: What are your strategies for adapting to change and leading teams through change?

Example response:
When it comes to adapting to change and leading teams through change, some strategies that I have found to be effective include:

1. *Communicating clearly:* It is important to communicate clearly and transparently with the team about any changes that are happening and the reasons for those changes. This helps to ensure that everyone is on the same page and that there is a shared understanding of the situation.
2. *Providing support:* It is also important to provide support to team members as they adapt to the change. This could involve offering training or resources to help them learn new skills or providing extra support during the transition period.
3. *Encouraging open communication:* Encouraging open communication within the team can help to create a sense of trust and foster a positive team culture. This could involve setting up regular check-ins or team meetings to discuss progress and any challenges that team members are facing.
4. *Being flexible:* Finally, it is important to be flexible and adapt to changing circumstances as they arise. This could involve revising plans or finding creative solutions to unexpected challenges.

In addition to the strategies I mentioned previously, there are also some technical strategies that can be helpful when adapting to change and leading teams through change:

1. *Using project management tools:* Project management tools, such as Trello or Asana, can help to keep track of tasks and progress, and can be particularly useful when adapting to change. These tools can help to ensure that everyone is aware of what needs to be done and when and can provide a clear overview of the project status.
2. *Establishing clear process and procedures:* Establishing clear process and procedures can help to ensure that everyone is on the same

page and that tasks are being completed efficiently. This could involve creating standard operating procedures or setting up workflow processes to guide the team's efforts.

3. *Using version control systems:* When working on software projects, using a version control system like Git can help to keep track of changes to the codebase and ensure that the team is working off the most up-to-date version. This can be especially useful when adapting to change and making updates to the code.

4. *Setting up automated testing and deployment pipelines:* Setting up automated testing and deployment pipelines can help to ensure that code changes are being thoroughly tested and that new features or updates are being deployed in a timely and reliable manner. This can be particularly helpful when adapting to change and making rapid updates to the product.

Q11: How do you approach managing the product roadmap and prioritizing features?

Example response:

I was the PM for a cloud-based project management tool that aimed to help teams collaborate more effectively and get work done more efficiently.

1. *Identified the goals and objectives of the product:* The first step was to identify the goals and objectives of the product. In this case, the goal was to help teams collaborate more effectively and get work done more efficiently.

2. *Gathered input from a variety of sources:* To ensure that the product was meeting the needs of the target audience, I gathered input from a variety of sources, such as current users of the tool,

industry experts, and internal stakeholders. This included conducting customer surveys, focus groups, and market research to gather insights and ideas.

3. *Defined the product vision and strategy:* Based on the input gathered, I defined the product vision and strategy. This involved outlining the long-term direction for the product and how the roadmap and features fit into that vision. For example, the vision was to create the most comprehensive and user-friendly project management tool on the market.

4. *Prioritized features:* Once the product vision and strategy had been defined, the next step was to prioritize the features that would be included on the roadmap. This involved using a framework such as the Kano model or the Jobs-to-be-Done framework to evaluate the value of each feature and prioritize them based on their importance to the target audience and alignment with the product vision. For example, features that were considered "must-have" were those that were essential for the product to fulfill its intended purpose, such as integration with popular productivity tools. "Nice-to-have" features, on the other hand, were those that would be beneficial but not necessary for the product to function, such as customizable dashboards or advanced reporting capabilities.

5. *Created the roadmap:* With the features prioritized, the next step was to create the roadmap. This involved outlining the timeline for delivering the features and any dependencies or constraints that needed to be considered. For example, the roadmap might have included a timeline for releasing new features on a quarterly basis, with specific milestones for each release.

6. *Reviewed and updated the roadmap regularly:* Finally, it was important to review and update the roadmap regularly to ensure that it remained relevant and aligned with the product vision and strategy. This involved revising the timeline or adding or removing features as needed based on customer feedback, market trends, and internal priorities. This helped to ensure that the product remained on track and aligned with the goals and objectives of the team.

Q12: What are your strategies for managing the budget and allocating resources for the product?

Example response:

As the product manager for a new mobile app, I was responsible for managing the budget and allocating resources for the development and release of the app.

One of the strategies I used was prioritization. I worked with the development team to identify the most important and impactful features that needed to be included in the initial release of the app. We used MoSCoW prioritization to rank these features, with the most essential features marked as "must-haves" and the least essential features marked as "nice-to-haves." This helped us to focus our resources on the most key features and ensure that the app was released on time.

I also used planning to carefully allocate resources for each feature. I used a Gantt chart to plan out the tasks and resources needed for each part of the development process, including budget, personnel, and other resources. I used planning poker to estimate the effort required for each task and worked with the development team to ensure that we had a clear understanding of the costs and risks associated with each feature.

To track progress against the budget and resource plan, I used project management software such as Asana or Jira. I regularly updated the status of each task and monitored the budget to ensure that we were staying on track and meeting our goals. I also used earned value analysis to compare the actual progress of the project against the planned progress and budget and identified any deviations or issues that needed to be addressed.

I communicated regularly with the team and other stakeholders to keep everyone informed about the budget and resource constraints, and to seek input and feedback on how best to allocate resources to achieve the desired outcomes. I also used online project management tools such as Trello or Slack to keep everyone up to date on the progress of the project and to facilitate communication.

Finally, I remained flexible and adaptable, and was willing to make changes to the budget and resource plan as needed in response to changing circumstances or priorities. This included adjusting the scope or timeline of the project, reassigning resources, or identifying alternative solutions to achieve the desired outcomes.

Q13: How do you go about measuring the success of a product and using data to inform future product decisions?

Example response:
As the product manager for a new e-commerce platform, I was responsible for measuring the success of the product and using data to inform future product decisions.

One of the key metrics I used to measure the success of the product was customer satisfaction. I conducted regular surveys to gather feedback from customers on their experience using the platform and used this data to identify areas for improvement. I also tracked key performance indicators (KPIs) such as conversion rates, average order value, and customer lifetime value to measure the financial performance of the product.

I used a variety of tools to collect and analyze data on the performance of the product. These included web analytics tools such as Google Analytics, customer relationship management (CRM) software, and customer feedback platforms such as UserTesting or SurveyMonkey.

Based on this data, I made informed decisions about the direction of the product and identified opportunities for growth and improvement. For example, if customer satisfaction was low in a certain area of the platform, I would work with the development team to prioritize improvements in that area. If the data showed that a particular feature was driving high levels of customer engagement and revenue, I might allocate more resources towards further development of that feature.

In addition to using data to inform product decisions, I also communicated regularly with the team and other stakeholders to share insights and progress and sought input and feedback on the direction of the product. This helped to ensure that we were making informed decisions that were aligned with the overall goals and objectives of the product.

Q14: What are your strategies for working with cross-functional teams to ensure that the product meets the needs of the customer?

Example response:

As the product manager for a new enterprise software platform, I was responsible for working with cross-functional teams to ensure that the product met the needs of the customer.

One of the strategies I used was effective communication. I worked closely with the development team, the design team, and other stakeholders to ensure that everyone had a clear understanding of the technical requirements and constraints of the product. I used tools such as Jira or Asana to track and coordinate the work of the various teams and held regular stand-up meetings and sprint planning sessions to ensure that everyone was on the same page.

Another strategy I used was user research and customer feedback. I conducted regular usability testing sessions and gathered feedback from customers through surveys and interviews. I also analyzed data from web analytics tools such as Google Analytics to identify patterns and trends in how users were interacting with the product. This data helped me to identify areas for improvement and prioritize features for development.

I also used agile development methodologies such as Scrum or Kanban to ensure that the product was delivered iteratively and incrementally, with regular feedback and review from the cross-functional teams. This helped to ensure that the product was continuously improving and meeting the needs of the customer.

Finally, I was flexible and adaptable, and was willing to make changes to the product roadmap as needed in response to changing circumstances or priorities. This included adjusting the scope or timeline of the project, reassigning resources, or identifying alternative solutions to achieve the desired outcomes. I used tools such as impact mapping or user story mapping to help visualize and plan these changes and worked with the development team to ensure that they were feasible and aligned with the overall goals and objectives of the product.

Q15: How do you approach managing the product lifecycle and ensuring that the product is successful in the market?

Example response:

In my previous role as a technical product manager at XYZ company, I was responsible for managing the product lifecycle for our cloud storage solution. To ensure the product was successful in the market, I used a combination of data analysis and customer feedback to inform my decision-making.

I regularly checked in with our sales team to understand how the product was performing in the market and gather feedback from customers. I also used tools like Google Analytics to track usage and engagement metrics, and A/B tested different features to see which ones resonated with our target audience.

In addition to tracking metrics, I also made it a priority to meet with key stakeholders and solicit their input on the product direction. This included conducting user research sessions and gathering feedback from our support team, who had direct interaction with customers.

By staying attuned to market trends and customer needs, we were able to pivot the product roadmap as needed and make data-driven decisions that led to the product's success in the market.

Q16: What are your strategies for building and maintaining relationships with customers and external stakeholders?

Example response:

When it comes to building and maintaining relationships with customers and external stakeholders, some strategies that I have found to be effective include:

1. *Communicate regularly:* One of the key strategies for building and maintaining relationships with customers and external stakeholders is to communicate regularly and keep them informed about the product and any updates or changes. This could involve sending out newsletters or email updates or setting up regular meetings or calls to discuss progress and address any concerns.

2. *Seek out feedback and listen to concerns:* It is also important to seek out feedback from customers and external stakeholders and listen to their concerns. This could involve conducting customer surveys or focus groups or having one-on-one conversations with key stakeholders to get their perspective. By gathering and acting on feedback, I can show that I value the input of customers and external stakeholders and build trust in the relationship.

3. *Address concerns and issues promptly:* When concerns or issues arise, it is important to address them promptly and proactively. This could involve providing solutions or alternatives or working with the team to resolve any issues as quickly as possible. By

showing that I am responsive and willing to address concerns, I can build trust and strengthen the relationship.

4. *Foster a sense of collaboration:* Building and maintaining relationships with customers and external stakeholders involves fostering a sense of collaboration and working together towards shared goals. This could involve sharing information or resources or seeking out opportunities to work together on projects or initiatives.

By following these strategies, I can build and maintain strong relationships with customers and external stakeholders, which is essential for the success of any product or business.

Q17: How do you approach managing the product launch process and ensuring a successful rollout?

Example response:
I was the PM for the launch of a new mobile app that aimed to help users track their fitness goals and progress.

1. *Defined the launch plan:* The first step in managing the product launch process was to define the launch plan. This involved outlining the steps that needed to be taken to prepare for the launch, such as conducting market research, developing marketing materials, and testing the product to ensure that it was ready for release. The launch plan included milestones for each phase of the process, as well as any dependencies or constraints that needed to be considered.

2. *Identified and assessed potential risks:* To ensure the success of the product launch, it was important to identify and assess

potential risks that could impact the rollout. This included conducting a risk assessment to identify potential issues that might arise during the launch process and developing contingency plans to mitigate those risks. For example, we identified the risk of potential technical issues with the app and put in place measures to test the app extensively and address any issues before the launch.

3. *Set up monitoring and tracking systems:* To track the performance of the app and identify any issues that needed to be addressed, we set up monitoring and tracking systems. This included setting up analytics tools to track user engagement and performance metrics, as well as conducting customer surveys to gather feedback. By regularly reviewing this data, we were able to identify any areas where the app was underperforming and make updates or changes as needed.

4. *Communicated with stakeholders:* Effective communication with stakeholders was critical for the success of the product launch. This involved keeping stakeholders informed about the launch plan and any updates or changes and seeking out their input and feedback throughout the process. We held regular meetings with the team and key stakeholders to discuss progress and address any concerns, and kept stakeholders informed through newsletters and email updates.

5. *Iterated and refined based on ongoing data gathering:* Finally, we were prepared to iterate and refine the app based on ongoing data gathering. This involved making updates or changes to the app based on user feedback or performance data, as well as revising the launch plan as needed. For example, based on customer feedback, we added a new feature to the app that allowed users to track their progress over time and set custom goals.

Q18: What are your strategies for managing the product roadmap in a fast-paced, constantly evolving market?

Example response:

As the product manager for a new social media platform, I was responsible for managing the product roadmap in a fast-paced, constantly evolving market.

One of the strategies I used was frequent and ongoing market research. I used a variety of tools and techniques to gather insights from customers, such as surveys, interviews, and usability testing. I also conducted regular competitive analysis to keep track of industry trends and the offerings of our competitors. I used tools such as Google Analytics and social media monitoring software to track metrics such as user engagement, retention, and conversion rates, and used this data to identify areas for improvement and opportunities for growth.

Another strategy I used was agile development methodologies such as Scrum or Kanban. I worked closely with the development team to define and prioritize user stories and used tools such as Jira or Asana to track and coordinate the work. I held regular stand-up meetings and sprint planning sessions to ensure that everyone was aligned on the goals and priorities of the project and used agile metrics such as velocity and burn-down charts to track progress and identify any issues that needed to be addressed.

I also used tools such as impact mapping or user story mapping to help visualize and plan the product roadmap and worked with the development team to prioritize and sequence the work. This helped to ensure that the most important and impactful features were delivered first, and that the product roadmap was aligned with the overall goals and objectives of the company.

Finally, I remained flexible and adaptable, and was willing to make changes to the product roadmap as needed in response to changing circumstances or priorities. This included adjusting the scope or timeline of the project, reassigning resources, or identifying alternative solutions to achieve the desired outcomes. I used agile techniques such as retrospectives or Lean Coffee to regularly review and revise the product roadmap and worked with the development team to ensure that these changes were feasible and aligned with the overall goals and objectives of the product.

Q19: How do you approach gathering and analyzing market and competitive data to inform product decisions?

Example response:

As the product manager for a new mobile app, I was responsible for gathering and analyzing market and competitive data to inform product decisions.

One of the key sources of data that I used was customer feedback. I conducted regular surveys and usability testing sessions to gather insights from users on their experience with the app and their needs and preferences. I also used social media monitoring tools such as Hootsuite or Brand24 to track mentions of the app on social media platforms and used this data to identify areas for improvement and opportunities for growth.

I also conducted regular competitive analysis to keep track of the offerings of our competitors and used tools such as SimilarWeb or App Annie to gather data on their performance and market share.

To make sense of this data, I used a variety of data visualization and analysis tools such as Excel or Tableau. I created charts and graphs to illustrate key trends and patterns and used statistical analysis tools such as R or SPSS to test hypotheses and draw more detailed conclusions about the data.

Based on this analysis, I made informed decisions about the direction of the product and identified opportunities for improvement and growth. For example, if customer satisfaction was low in a certain area of the app, I might work with the development team to prioritize improvements in that area. If the data showed that a particular feature was particularly popular among users, I might allocate more resources towards further development of that feature.

I communicated the results of the data analysis to the relevant stakeholders, including the development team and other decision-makers, using clear and concise presentations and reports. I sought input and feedback from these stakeholders on how the data should inform product decisions and incorporated this feedback into the product roadmap.

Q20: What are your strategies for managing the product roadmap and aligning it with the overall business strategy?

Example response:

As the product manager for a new e-commerce platform, I was responsible for managing the product roadmap and aligning it with the overall business strategy.

To start, I worked closely with senior management to understand the long-term vision for the company and the role that the product played in achieving this vision. Based on this understanding, I defined clear product goals and objectives that aligned with the overall business strategy. These included revenue targets for the first year, as well as metrics such as user growth and customer satisfaction.

I used tools such as impact mapping or user story mapping to visualize and plan the product roadmap and worked with the development team to prioritize and sequence the work. I used agile development methodologies such as Scrum or Kanban to ensure that the product was delivered iteratively and incrementally, with regular feedback and review from the development team and other stakeholders.

I regularly reviewed and revised the product roadmap to ensure that it remained aligned with the overall business strategy and the needs and expectations of the customer. This included using agile techniques such as retrospectives or Lean Coffee to identify and address any issues or challenges and adjusting the scope or timeline of the project as needed.

I communicated and coordinated with relevant stakeholders, including the development team, sales and marketing teams, and senior management, to ensure that everyone was aligned on the product roadmap and the overall business strategy. I used tools such as Jira or Asana to track and coordinate the work and held regular meetings and progress reviews to ensure that everyone was on the same page.

As a result of these efforts, the product was successfully launched and achieved the revenue and user growth targets that had been set. The product roadmap remained aligned with the overall business strategy, and the product was well-received by the customer.

Q21: What are your strategies for managing the product lifecycle and ensuring that the product remains relevant in the market?

Example response:

As the product manager for a software tool used by HR professionals, I was responsible for managing the product lifecycle and ensuring that the product remained relevant in the market.

One of the strategies I used was ongoing market research. I regularly gathered insights from customers, industry trends, and competitors to understand the evolving needs and expectations of the target audience. I used tools such as surveys, interviews, and usability testing to gather this data, and analyzed it using data visualization and analysis tools such as Excel or Tableau.

Based on this analysis, I identified opportunities for improvement and growth, and used agile development methodologies such as Scrum or Kanban to deliver new features and updates iteratively and incrementally to the product. I worked closely with the development team to define and prioritize user stories and used tools such as Jira or Asana to track and coordinate the work.

I also used tools such as impact mapping or user story mapping to help visualize and plan the product roadmap and worked with the development team to prioritize and sequence the work. This helped to ensure that the most important and impactful features were delivered first, and that the product roadmap was aligned with the overall goals and objectives of the company.

Finally, I remained flexible and adaptable, and was willing to make changes to the product roadmap as needed in response to changing circumstances or priorities. This included adjusting the scope or timeline of the project, reassigning resources, or identifying alternative solutions to achieve the desired outcomes. I used agile techniques such as retrospectives or Lean Coffee to regularly review and revise the product roadmap and worked with the development team to ensure that these changes were feasible and aligned with the overall goals and objectives of the product.

Through these efforts, I was able to successfully manage the product lifecycle and ensure that the product remained relevant and competitive in the market.

Q22: What are your strategies for managing the product lifecycle and ensuring that the product remains relevant in the market?

Example response:

As the product manager for a social media platform, I was responsible for gathering and analyzing customer feedback to continuously improve the product.

To gather customer feedback, I used a variety of tools and techniques including online surveys, usability testing sessions, and social media monitoring. For example, I used Qualtrics to design and administer surveys to gather insights on users' experience with the platform, and UserTesting to conduct usability testing sessions to identify problems and areas for improvement. I also used tools such as Hootsuite or Brand24 to track mentions of the platform on social media platforms and gather customer feedback from these sources.

I analyzed the results of these efforts using data visualization and analysis tools such as Excel or Tableau. I created charts and graphs to illustrate key trends and patterns and used statistical analysis tools such as R or SPSS to test hypotheses and draw more detailed conclusions about the data.

Based on this analysis, I identified areas for improvement and worked with the development team to prioritize and implement updates and new features. I used agile development methodologies such as Scrum or Kanban to ensure that the updates were delivered iteratively and incrementally, with regular feedback and review from the development team and other stakeholders.

I also regularly solicited feedback from customers through in-app surveys, emails, or online forums, and used this feedback to inform my decisions about the direction of the product.

Finally, I communicated the results of the customer feedback analysis to the relevant stakeholders, including the development team and other decision-makers, using clear and concise presentations and reports. I sought input and feedback from these stakeholders on how the data should inform product decisions and incorporated this feedback into the product roadmap using tools such as Jira or Asana to track and coordinate the work and held regular meetings and progress reviews to ensure that everyone was on the same page.

Through these efforts, I was able to continuously gather and analyze customer feedback and use this information to drive improvements to the product. I was able to identify key trends and patterns in the data and use statistical analysis tools to test hypotheses and draw more detailed conclusions. I worked closely with the development team to deliver updates and new features that addressed the needs and preferences of the customer and used agile development methodologies to ensure that these updates were delivered iteratively and incrementally. I also sought input and feedback from stakeholders and incorporated this feedback into the product roadmap, ensuring that the product remained relevant and competitive in the market.

Q23: What are your strategies for managing the product roadmap and adapting to changing market conditions?

Example response:

As the product manager for a mobile app, I was responsible for managing the product roadmap and adapting to changing market conditions.

To inform product decisions, I regularly gathered and analyzed market and competitive data using a variety of tools and techniques. For example, I used Google Analytics to track the performance of the app in the market and conducted research on industry trends and competitors using sources such as industry publications or analysts' reports. I also used data visualization and analysis tools such as Excel or Tableau to create charts and graphs illustrating key trends and patterns and used statistical analysis tools such as R or SPSS to test hypotheses and draw more detailed conclusions about the data.

Based on this analysis, I identified opportunities for improvement and growth, and used agile development methodologies such as Scrum or Kanban to deliver new features and updates iteratively and incrementally to the app. I worked closely with the development team to define and prioritize user stories and used tools such as Jira or Asana to track and coordinate the work.

I also used tools such as impact mapping or user story mapping to help visualize and plan the product roadmap and worked with the development team to prioritize and sequence the work. This helped to ensure that the most important and impactful features were delivered first, and that the product roadmap was aligned with the overall goals and objectives of the company.

Finally, I remained flexible and adaptable, and was willing to make changes to the product roadmap as needed in response to changing market conditions or priorities. This included adjusting the scope or timeline of the project, reassigning resources, or identifying alternative solutions to achieve the desired outcomes. I used agile techniques such as retrospectives or Lean Coffee to regularly review and revise the product roadmap and worked with the development team to ensure that these changes were feasible and aligned with the overall goals and objectives of the product.

Through these efforts, I was able to effectively manage the product roadmap and adapt to changing market conditions, ensuring that the app remained relevant and competitive in the market.

Q24: How do you approach building and maintaining relationships with cross-functional teams and ensuring that the product meets the needs of all stakeholders?

Example response:

As the product manager for a software platform, I was responsible for building and maintaining relationships with cross-functional teams and ensuring that the product met the needs of all stakeholders.

One of the key strategies I used was to establish clear communication channels and processes with the development team, sales team, and other stakeholders. I held regular meetings with these teams to discuss the status of the product, gather feedback, and coordinate efforts. I also used tools such as Slack or Microsoft Teams to facilitate ongoing communication and collaboration and used project management tools such as Jira or Asana to track and coordinate the work.

I also worked to establish trust and build relationships with the development team and other stakeholders by demonstrating my technical knowledge and expertise, and by being transparent and open about my decision-making process. I sought input and feedback from these stakeholders and tried to understand their needs and perspectives.

In addition, I used agile development methodologies such as Scrum or Kanban to ensure that the product was delivered iteratively and incrementally, with regular feedback and review from the development team and other stakeholders. This helped to ensure that the product met the needs of all stakeholders and was aligned with the overall goals and objectives of the company.

Finally, I remained flexible and adaptable, and was willing to make changes to the product roadmap as needed in response to changing circumstances or priorities. This included adjusting the scope or timeline of the project, reassigning resources, or identifying alternative solutions to achieve the desired outcomes. I used agile techniques such as retrospectives or Lean Coffee to regularly review and revise the product roadmap and worked with the development team and other stakeholders to ensure that these changes were feasible and aligned with the overall goals and objectives of the product.

Through these efforts, I was able to effectively build and maintain relationships with cross-functional teams and ensure that the product met the needs of all stakeholders. I used clear communication channels and processes and sought input and feedback from these stakeholders to understand their needs and perspectives. I used agile development methodologies to ensure that the product was delivered iteratively and incrementally and remained flexible and adaptable to changing circumstances or priorities. As a result, I was able to deliver a product that was aligned with the overall goals and objectives of the company and met the needs of all stakeholders.

Q25: How do you use data analysis and visualization tools such as Excel, Tableau, or R to inform product decisions, and can you provide an example of a specific problem you solved using these tools?

Example response:

As the product manager for an e-commerce platform, I was responsible for using data analysis and visualization tools such as Excel, Tableau, or R to inform product decisions.

One specific problem that I solved using these tools was to identify patterns in customer behavior and improve the recommendation algorithm for our product recommendation engine.

To solve this problem, I first gathered and cleaned the relevant data using tools such as SQL or Python. This data included customer purchase history, product metadata, and other relevant variables such as customer demographics or location.

Next, I used tools such as Excel or Tableau to create charts and graphs illustrating key trends and patterns in the data. For example, I created scatter plots to visualize the relationship between different variables or used pivot tables to summarize and group the data in different ways.

Finally, I used statistical analysis tools such as R or SPSS to test hypotheses and draw more detailed conclusions about the data. For example, I used regression analysis to identify the key variables that were most predictive of customer behavior or used clustering algorithms to identify different segments of customers with similar characteristics.

Based on these analyses, I was able to identify patterns in customer behavior that were not immediately apparent from the raw data and used these insights to improve the recommendation algorithm for our product recommendation engine. This resulted in a significant increase in customer engagement and satisfaction and contributed to the overall growth of the business.

CHAPTER 6
Case Questions Answers and Explanations

Q1: You are the PM for a new subscription-based meal delivery service launching in a highly competitive market. What steps would you take to differentiate your service from competitors and ensure its success?

Example response:

As the PM for a new subscription-based meal delivery service launching in a highly competitive market, it is important to take a proactive and data-driven approach to differentiate the service from competitors and ensure its success. Here are some steps that could be taken:

1. *Research the market:* The first step in launching a successful meal delivery service is to thoroughly research the market. This may involve analyzing market trends and demographics, understanding the competitive landscape, and identifying the needs and preferences of the target audience.

2. *Develop a unique value proposition:* Based on the research, develop a unique value proposition that sets the service apart from competitors. This may involve offering a unique product or service, such as a specialized meal plan or a proprietary delivery method, or it may involve offering a superior customer experience, such as exceptional customer service or convenient delivery options.

3. *Gather and analyze data:* To effectively differentiate the service and ensure its success, it is important to gather and analyze data on a regular basis. Some data sources that may be helpful to consider include:

    a. *Customer data:* Analyzing customer data, such as demographics, purchasing behavior, and feedback, can help to identify trends and opportunities for improvement or differentiation.

    b. *Usage data:* Analyzing usage data, such as the number of orders placed, the average order value, and the frequency of use, can help to identify trends in customer behavior and identify areas for optimization.

    c. *Market data:* Analyzing market data, such as market trends and competitive analysis, can help to identify opportunities for differentiation and inform the development of the marketing strategy.

To analyze this data, a variety of techniques can be used, including:

1. *Data visualization:* Visualizing the data in charts or graphs can help to identify trends and patterns in the data and can make it easier to understand and communicate the findings.

2. *Statistical analysis:* Using statistical techniques such as regression analysis or cluster analysis can help to identify relationships and trends in the data that may not be immediately apparent.

3. *Qualitative analysis:* Analyzing open-ended responses from surveys or focus groups can provide insights into the attitudes and motivations of customers and can help to identify opportunities for improvement or differentiation.

4. *Implement and test changes:* Based on the data gathered and analyzed, implement changes to the service to differentiate it from competitors and improve its success. This may involve adjusting the product or service offering, the pricing, the marketing, or the customer experience. It is important to test these changes to ensure that they are effective in improving the service and meeting the needs of customers.

5. *Monitor and respond to customer feedback:* It is important to continue to monitor and respond to customer feedback to ensure that the meal delivery service meets the needs of customers. This may involve implementing additional features or functionality based on customer requests or addressing any issues or concerns that are raised by customers.

6. *Continuously gather and analyze data:* To continue to optimize the meal delivery service for success, it is important to continuously gather and analyze data. This may involve tracking key performance indicators such as customer satisfaction, retention, and revenue, and using the data to inform ongoing improvements to the service.

7. *Seek external expertise:* If the market is particularly complex or the internal team lacks the necessary expertise to succeed in the competitive market, it may be helpful to seek external

expertise. This could involve consulting with industry experts or partnering with external companies or organizations that have relevant expertise.

Overall, by gathering and analyzing data, developing a unique value proposition, and continuously monitoring and responding to customer feedback, the PM can differentiate the meal delivery service from competitors and increase the chances of success in a highly competitive market.

Q2: You are the PM for an e-commerce platform that is launching in a new market. What steps would you take to ensure that the platform is successful in this new market?

Example response:

To ensure success in a new market, it is important to gather and analyze a variety of data sources to understand the needs and preferences of the target audience. Some data sources that may be helpful to consider include:

1. *Market trends and demographics:* Analyzing market trends and demographics can help to understand the size and characteristics of the potential customer base in the new market. This data may be available through public sources such as government statistics agencies or industry research firms, or it may be collected through primary research methods such as surveys or focus groups.

2. *Competitive landscape:* Understanding the competitive landscape can help to identify opportunities for differentiation and position the e-commerce platform in a way that is unique and appealing

to the target audience. This data may be gathered through market research firms, industry publications, or by conducting primary research such as customer interviews or surveys.

3. *User feedback:* Gathering feedback from users through surveys, support tickets, and other channels can provide valuable insights into what users like and dislike about the e-commerce platform, as well as their needs and preferences in the new market.

To analyze this data, a variety of techniques can be used, including:

1. *Data visualization:* Visualizing the data in charts or graphs can help to identify trends and patterns in the data and can make it easier to understand and communicate the findings.
2. *Statistical analysis:* Using statistical techniques such as regression analysis or cluster analysis can help to identify relationships and trends in the data that may not be immediately apparent.
3. *Qualitative analysis:* Analyzing open-ended responses from surveys or focus groups can provide insights into the attitudes and motivations of the target audience and can help to identify opportunities for improvement or differentiation.

To ensure success in the new market, the following steps can be taken:

1. *Use the data to inform the market-specific strategy:* Use the data gathered and analyzed to inform the development of a market-specific strategy that considers the unique needs and preferences of the target audience in the new market. This may involve adjusting the product offering, pricing, marketing, or customer service to better meet the needs of the target audience.
2. *Implement technical solutions to address market-specific challenges:* Based on the data and the market-specific strategy, implement

technical solutions to address any challenges that may be specific to the new market. This may involve integrating with local payment gateways or shipping carriers or adapting the platform to meet local regulatory or compliance requirements.

3. *Monitor and respond to user feedback:* It is important to continue to monitor and respond to user feedback to ensure that the e-commerce platform meets the needs of the target audience in the new market. This may involve implementing additional features or functionality based on user requests or addressing any issues or concerns that are raised by users.

4. *Continuously gather and analyze data:* To continue to optimize the e-commerce platform for success in the new market, it is important to continuously gather and analyze data. This may involve tracking key performance indicators such as conversion rates, average order value, and customer retention, and using the data to inform ongoing improvements to the platform.

5. *Seek external expertise:* If the market is particularly complex or the internal team lacks the necessary expertise to succeed in the new market, it may be helpful to seek external expertise. This could involve consulting with industry experts or partnering with external companies or organizations that have relevant expertise.

Q3: You are the PM for a social media platform that has seen a decline in user engagement over the past year. What steps would you take to identify the root cause of the decline and implement a plan to increase engagement?

Example response:

As the PM for a social media platform that has seen a decline in user engagement over the past year, it is important to take a proactive and data-driven approach to identifying the root cause of the decline and implementing a plan to increase engagement. Here are some steps that could be taken:

1. *Gather and analyze data:* To effectively identify the root cause of the decline in user engagement, it is important to gather and analyze data on a regular basis. Some data sources that may be helpful to consider include:

    a. *Engagement data:* Analyzing engagement data, such as the number of likes, comments, and shares on posts, can help to identify trends in user behavior and identify areas for optimization.

    b. *Usage data:* Analyzing usage data, such as the average time spent on the platform and the number of daily active users, can help to identify trends in user behavior and identify areas for optimization.

    c. *Customer data:* Analyzing customer data, such as demographics, interests, and feedback, can help to identify trends and opportunities for improvement or differentiation.

To analyze this data, a variety of techniques can be used, including:

1. *Data visualization:* Visualizing the data in charts or graphs can help to identify trends and patterns in the data and can make it easier to understand and communicate the findings.

2. *Statistical analysis:* Using statistical techniques such as regression analysis or cluster analysis can help to identify relationships and trends in the data that may not be immediately apparent.

3. *Qualitative analysis:* Analyzing open-ended responses from surveys or focus groups can provide insights into the attitudes and motivations of users and can help to identify opportunities for improvement or differentiation.

4. *Use technical tools:* There are a variety of technical tools that can be used to gather and analyze data and inform the development of a plan to increase engagement. For example, analytics tools can help to track and analyze user behavior on the platform, while customer relationship management (CRM) systems can help to track and manage customer interactions.

5. *Develop a plan:* Based on the data gathered and analyzed, develop a plan to increase engagement on the social media platform. This may involve implementing new features or functionality, adjusting the content strategy, or improving the customer experience.

6. *Test and iterate:* Before launching any changes, it is important to test them to ensure that they effectively address the issue of declining engagement. This may involve conducting user testing, A/B testing, or other forms of experimentation. If necessary, continue to iterate and make further changes until the desired increase in engagement is achieved.

7. *Monitor and respond to customer feedback:* It is important to continue to monitor and respond to customer feedback to ensure that the social media platform meets the needs of users and continues to drive engagement. This may involve implementing additional features or functionality based on customer requests or addressing any issues or concerns that are raised by users.

8. *Continuously gather and analyze data:* To continue to optimize the social media platform for success and drive engagement, it is important to continuously gather and analyze data. This may involve tracking key performance indicators such as user engagement, retention, and satisfaction, and using the data to inform ongoing improvements to the platform.

Q4: **You are the PM for an online marketplace that is struggling to monetize. What steps would you take to identify new monetization opportunities and implement them effectively?**

Example response:

As the PM for an online marketplace that is struggling to monetize, it is crucial to take a proactive and data-driven approach to identifying new monetization opportunities and implementing them effectively.

1. *Gather and analyze data:* To identify new monetization opportunities, it is essential to gather and analyze data regularly. This could include revenue data (e.g., total revenue generated and average transaction value), customer data (e.g., demographics, purchasing behavior, and feedback), and usage data (e.g., number of orders placed, average time spent on the platform, and number of daily active users). Data visualization, statistical analysis, and qualitative analysis techniques can be used to identify trends and opportunities for optimization or differentiation.

2. *Utilize technical tools:* Technical tools, such as analytics and customer relationship management systems, can be used to gather and analyze data and inform the identification of new monetization opportunities.

3. *Identify new monetization opportunities:* Based on the data gathered and analyzed, identify new monetization opportunities that could be implemented on the online marketplace. This may involve introducing new products or services, adjusting the pricing model, or implementing new features or functionality.

4. *Test and iterate:* Before launching any changes, test them to ensure that they effectively address the monetization issue. This may involve user testing, A/B testing, or other forms of experimentation. If needed, continue to iterate, and make additional changes until the desired increase in monetization is achieved.

5. *Monitor and respond to customer feedback:* It is essential to continue to monitor and respond to customer feedback to ensure that the online marketplace meets the needs of customers and continues to drive monetization. This may involve implementing additional features or functionality based on customer requests or addressing any issues or concerns that are raised by customers.

6. *Continuously gather and analyze data:* To continue to optimize the online marketplace for success and drive monetization, it is essential to continuously gather and analyze data. This may involve tracking key performance indicators such as customer satisfaction, retention, and revenue, and using the data to inform ongoing improvements to the platform.

By gathering and analyzing data and utilizing technical tools, the PM can effectively identify new monetization opportunities and implement them effectively on the online marketplace. This approach can help to improve the customer experience and drive monetization on the platform.

Q5: You are the PM for a messaging app that has seen a decline in usage over the past few months. What steps would you take to understand the cause of the decline and implement a plan to increase usage and engagement?

Example response:

To increase engagement of a messaging app, it would be helpful to have access to a variety of data sources, including:

1. *Usage data:* This could include metrics such as active user counts, retention rates, session lengths, and frequency of use. This data can help to identify trends in usage over time and can provide insights into what features or functionality are most popular with users.
2. *User feedback:* Gathering feedback from users through surveys, support tickets, and other channels can provide valuable insights into what users like and dislike about the app, as well as their needs and expectations.
3. *Market trends:* Understanding trends in the broader market, such as new features or functionality that are becoming popular, can help to inform the development of new features or functionality for the app.
4. *Technical performance data:* This could include metrics such as server uptime, page load times, and error rates. This data can help to identify technical issues that may be impacting usage and engagement and can inform the development of technical solutions to address those issues.

To use this data to increase engagement, the following steps could be taken:

1. *Analyze the data:* Begin by analyzing the data to identify trends and patterns. This may involve visualizing the data in charts or graphs or using statistical analysis techniques to identify trends or correlations.

2. *Identify opportunities for improvement:* Based on the analysis of the data, identify opportunities for improvement in the app. This may involve implementing new features or functionality that are in high demand or addressing technical issues that are causing user frustration or dissatisfaction.

3. *Develop a plan to address the opportunities:* Develop a plan to address the opportunities for improvement identified in the data. This may involve working with the development team to implement new features or functionality or implementing technical solutions to address technical issues.

4. *Test and iterate:* Before launching any changes, it is important to test them to ensure that they effectively address the issues identified in the data. This may involve conducting user testing, A/B testing, or other forms of experimentation. If necessary, continue to iterate and make further changes until the desired increase in usage and engagement is achieved.

5. *Monitor and respond to user feedback:* Once the changes have been launched, it is important to continue to monitor and respond to user feedback to ensure that the app meets the needs of the target audience. This may involve implementing additional features or functionality based on user requests or addressing any issues or concerns that are raised by users.

Q6: You are the PM for a ride-sharing app that is experiencing high churn rates among drivers. What steps would you take to reduce churn and improve driver retention?

Example response:

To effectively reduce churn and improve driver retention, it is important to gather and analyze a variety of data sources to understand the root cause of the high churn rates. Some data sources that may be helpful to consider include:

1. *Churn data:* Analyzing churn data can help to identify trends in driver behavior, such as when drivers are most likely to leave the app and what factors may be contributing to their decision to leave. This data may be available through the ride-sharing app's internal systems, or it may be collected through surveys or focus groups with drivers who have left the app.

2. *Driver feedback:* Gathering feedback from drivers through surveys, support tickets, and other channels can provide valuable insights into what drivers like and dislike about the app, as well as their needs and preferences.

3. *Usage data:* Analyzing usage data, such as the number of rides completed by drivers, the average length of a driver's shift, and the number of support tickets submitted by drivers, can help to identify trends in driver behavior and identify areas for improvement.

To analyze this data, a variety of techniques can be used, including:

1. *Data visualization:* Visualizing the data in charts or graphs can help to identify trends and patterns in the data and can make it easier to understand and communicate the findings.

2. *Statistical analysis:* Using statistical techniques such as regression analysis or cluster analysis can help to identify relationships and trends in the data that may not be immediately apparent.
3. *Qualitative analysis:* Analyzing open-ended responses from surveys or focus groups can provide insights into the attitudes and motivations of drivers and can help to identify opportunities for improvement or differentiation.

Based on the analysis of the data, the following steps can be taken to reduce churn and improve driver retention:

1. *Develop a plan to address the root cause:* Use the data gathered and analyzed to inform the development of a plan to address the root cause of the high churn rates. This may involve implementing new features or functionality to improve the driver experience, adjusting the pay structure, or improving support and communication with drivers.
2. *Test and iterate:* Before launching any changes, it is important to test them to ensure that they effectively address the issue. This may involve conducting user testing, A/B testing, or other forms of experimentation. If necessary, continue to iterate and make further changes until the desired reduction in churn is achieved.
3. *Monitor and respond to driver feedback:* It is important to continue to monitor and respond to driver feedback to ensure that the ride-sharing app meets the needs of drivers. This may involve implementing additional features or functionality based on driver requests or addressing any issues or concerns that are raised by drivers.
4. *Continuously gather and analyze data:* To continue to optimize the app for improved driver retention, it is important to continuously

gather and analyze data. This may involve tracking key performance indicators such as churn rates, driver satisfaction, and the number of support tickets submitted, and using the data to inform ongoing improvements to the app.

5. *Seek external expertise:* If the internal team lacks the necessary expertise to effectively address the issue of high churn rates, it may be helpful to seek external expertise. This could involve consulting with industry experts or partnering with external companies or organizations that have relevant expertise.

6. It is also important to consider implementing additional initiatives to improve driver retention, such as offering incentives or rewards programs for drivers who meet certain performance milestones or providing additional support and resources to help drivers succeed on the platform. Overall, a comprehensive and data-driven approach, combined with ongoing engagement and support for drivers, can help to effectively reduce churn and improve driver retention for the ride-sharing app.

Q7: Your company's existing video streaming service has low user adoption and usage. How would you design a new feature for the service that increases user adoption and usage?

Example response:

To design a new feature for a video streaming service that increases user adoption and usage, it is important to take a user-centered approach and consider the needs and preferences of the target audience. Here are some steps that could be taken:

1. *Research the target audience:* To design a new feature that effectively meets the needs of the target audience, it is important to

understand their behaviors, needs, and preferences. This may involve conducting user research, such as focus groups, surveys, or usability testing, to gather insights into how users currently use the video streaming service and what they are looking for in a new feature.

2. *Define the problem:* Based on the research conducted, define the specific problem that the new feature is intended to solve. This may involve identifying pain points or areas for improvement in the user experience or identifying opportunities to differentiate the service from competitors.

3. *Ideate solutions:* Generate a range of potential solutions to the problem identified. This may involve brainstorming sessions with a cross-functional team or gathering ideas from other sources such as customer feedback or industry best practices.

An example solution would be to develop a personalized content recommendation system that uses machine learning algorithms to recommend content based on the user's viewing history, ratings, and preferences.

To implement this solution, the following steps could be taken:

1. *Gather data:* To develop a personalized content recommendation system, it is necessary to gather data on user viewing history, ratings, and preferences. This data can be collected through tracking user interactions with the video streaming service and storing it in a data warehouse.

2. *Train a machine learning model:* Using the data gathered, train a machine learning model to predict which content a user is likely to enjoy based on their viewing history, ratings, and preferences.

This could be done using techniques such as collaborative filtering or matrix factorization.

3. *Implement the recommendation system:* Implement the recommendation system on the video streaming service, using the trained machine learning model to generate personalized recommendations for each user. This may involve integrating the recommendation system into the user interface or developing APIs to enable the recommendation system to be accessed by other applications.

4. *Monitor and optimize the recommendation system:* Monitor the performance of the recommendation system, using metrics such as click-through rate and retention to measure its effectiveness. Continuously gather and analyze data to optimize the recommendation system and improve its accuracy.

Overall, by developing a personalized content recommendation system using machine learning, it is possible to increase user adoption and usage of the video streaming service by providing users with personalized recommendations that are tailored to their interests and preferences.

Q8: **You are the PM for a new social media platform that is launching in the next few months. You have just received feedback from a focus group indicating that the platform is not meeting the needs of the target audience. What steps would you take to address this issue before the launch?**

Example response:

As the PM for a new social media platform, it is essential to gather and consider feedback from focus groups to ensure that the platform meets the needs of the target audience. In this case, receiving feedback indicating that the platform is not meeting the needs of the target audience is cause for concern and requires immediate action. Here are some steps I would take to address this issue before the launch:

1. *Review the feedback carefully:* It is important to thoroughly review the feedback from the focus group to understand the specific technical issues that are causing the platform to not meet the needs of the target audience. This may involve reviewing written comments or transcripts of focus group discussions, as well as analyzing log data and metrics to identify areas of the platform that are causing user frustration or dissatisfaction.

2. *Identify the root cause of the problem:* Once the specific technical issues have been identified, it is important to determine the root cause of the problem. This may involve asking additional questions of the focus group or conducting further research to understand why the platform is not meeting the needs of the target audience. It may also involve working with the development team to debug and troubleshoot technical issues, or consulting with subject matter experts to identify potential solutions.

3. *Develop a plan to address the issue:* Based on the root cause of the problem, develop a technical plan to address the issue before the launch. This may involve making changes to the platform's codebase, database schema, or infrastructure to improve the platform's performance, reliability, or scalability. It may also involve implementing new features or functionality to better meet the needs of the target audience.

4. *Test and iterate:* Before launching the platform, it is important to test the changes that have been made to ensure that they

effectively address the technical issues identified by the focus group. This may involve conducting unit tests, integration tests, and user acceptance tests to validate the changes. If necessary, continue to iterate and make further changes until the platform meets the technical and functional requirements of the target audience.

5. *Communicate with stakeholders:* Throughout this process, it is important to keep stakeholders informed of the technical steps being taken to address the issue and the progress that has been made. This may include the focus group, the development team, and any other relevant parties. It may also involve updating technical documentation and providing training to relevant personnel to ensure that they are prepared to support the platform upon launch.

6. *Gather additional feedback:* In addition to the feedback from the focus group, it may be helpful to gather additional feedback from other sources, such as user testing, surveys, or customer interviews. This can provide a more comprehensive understanding of the needs and expectations of the target audience and may identify additional issues that need to be addressed.

7. *Engage with the community:* Engaging with the community can help to build trust and establish a rapport with potential users of the platform. This may involve participating in online forums or social media groups related to the platform's target audience, or hosting events or webinars to interact with users in real-time.

8. *Monitor and respond to user feedback post-launch:* Once the platform is launched, it is important to continue to monitor and respond to user feedback to ensure that the platform continues to meet the needs of the target audience. This may involve implementing

additional features or functionality based on user requests or addressing any issues or concerns that are raised by users.

9. *Evaluate and refine the marketing strategy:* If the feedback indicates that the platform is not meeting the needs of the target audience because it is not being effectively marketed to them, it may be necessary to evaluate and refine the marketing strategy. This could involve adjusting the messaging or positioning of the platform or targeting different segments of the market.

10. *Seek external expertise:* If the issue is particularly complex or the internal team lacks the necessary expertise to address it, it may be helpful to seek external expertise. This could involve consulting with industry experts or partnering with external companies or organizations that have relevant expertise.

Q9: **You are the PM for a streaming video platform that is launching in a new country. What steps would you take to understand the preferences and expectations of the target market in this country, and how would you tailor the platform to meet those needs?**

Example response:

To implement this solution, the following steps could be taken:

1. *Identify the key questions to ask:* Identify the key questions that need to be asked to understand the preferences and expectations of users in the new country. These questions might include:

    a. What types of content do users in this country prefer to watch?

b. How do users in this country prefer to access content (e.g., through a web browser, mobile app, TV)?

   c. What are the key features and functionality that users in this country expect from a streaming video platform?

   d. What are the main pain points or areas for improvement that users in this country experience when using a streaming video platform?

2. *Design the survey tool:* Design the survey tool, including the questions that will be asked and the user interface. Consider factors such as ease of use, clarity of language, and the types of response options that will be provided.

3. *Integrate the survey tool into the streaming video platform:* Integrate the survey tool into the streaming video platform, either as a standalone feature or as part of the user onboarding process. Consider the optimal timing and frequency for conducting surveys to gather the most relevant and useful data.

4. *Gather and analyze data:* Use the survey tool to gather data from users in the new country and analyze the data to understand the preferences and expectations of the target market. This may involve using techniques such as data visualization or statistical analysis to identify trends and patterns in the data.

5. *Tailor the platform to meet the needs of the target market:* Based on the data gathered and analyzed, make changes to the streaming video platform to better meet the needs and expectations of users in the new country. This may involve adjusting the content offering, adding, or modifying features and functionality, or adjusting the user interface.

Q10: You are the PM for a home security app that is launching in a new country. What steps would you take to research the market, understand the needs and preferences of the target audience, and tailor the app to meet those needs?

Example response:

To research the market, understand the needs and preferences of the target audience, and tailor a home security app to meet those needs in a new country, the following steps could be taken:

1. *Research the market:* Conduct market research to understand the competitive landscape, regulatory environment, and overall demand for home security solutions in the new country. This may involve gathering data on the size and growth of the market, the types of products and services offered by competitors, and the key trends and challenges in the market.

2. *Identify the target audience:* Identify the target audience for the home security app in the new country. This may involve defining the demographics, behaviors, and needs of the target audience based on the market research conducted.

3. *Gather insights from the target audience:* Gather insights from the target audience to understand their needs and preferences in relation to home security. This may involve conducting focus groups, surveys, or usability testing to gather feedback and insights from potential users.

4. *Define the product roadmap:* Based on the insights gathered from the target audience, define the product roadmap for the home security app in the new country. This may involve identifying key features and functionality that need to be included in the

app, as well as any localizations or customization that need to be made to meet the needs and preferences of the target audience.

5. *Develop the app:* Develop the home security app based on the product roadmap defined. This may involve working with a cross-functional team of developers, designers, and subject matter experts to build the app and ensure that it meets the needs and preferences of the target audience in the new country.

6. *Test and optimize:* Conduct user testing and gather feedback on the home security app to identify areas for improvement. Use this feedback to optimize the app and ensure that it meets the needs and preferences of the target audience.

7. *Launch and monitor:* Once the home security app is developed and optimized, launch it in the new country and monitor its performance. This may involve tracking key performance indicators such as user adoption, retention, and satisfaction, and using the data to inform ongoing improvements to the app.

To develop the app, the following steps should be taken:

1. *Define the scope and goals of the app:* Define the scope and goals of the app, including the key features and functionality that will be included and the target audience that the app is intended to serve.

2. *Assemble a cross-functional team:* Assemble a cross-functional team of developers, designers, and subject matter experts who will be responsible for building the app. This team should have the skills and expertise needed to develop the app in accordance with the scope and goals defined.

3. *Define the user journey:* Define the user journey for the app, including the steps that users will take to accomplish their goals and the key interactions and decision points that will be included.

4. *Create wireframes and prototypes:* Create wireframes and prototypes of the app to define the layout and functionality of the app. These wireframes and prototypes can be used to test and validate the design of the app with users and stakeholders.

5. *Develop the app:* Develop the app using an agile development process, with a focus on rapid iteration and continuous delivery. This may involve building out the user interface, implementing the core functionality of the app, and integrating any necessary APIs or third-party services.

6. *Test and optimize:* Conduct user testing and gather feedback on the app to identify areas for improvement. Use this feedback to optimize the app and ensure that it meets the needs and preferences of the target audience.

7. *Launch and monitor:* Once the app is developed and optimized, launch it, and monitor its performance. This may involve tracking key performance indicators such as user adoption, retention, and satisfaction, and using the data to inform ongoing improvements to the app.

CHAPTER 7
Obscure Questions Answers and Explanations

Q1: If you were a character in a book or movie, who would you be and why?

Example response:
If I were a character in a book or movie, I would choose to be Atticus Finch from *To Kill a Mockingbird*. Atticus is a highly respected lawyer who is known for his intelligence, integrity, and commitment to justice. He can analyze complex situations and make logical, well-reasoned decisions, even in the face of difficult challenges.

I believe these qualities would be particularly valuable in a technical PM role, where the ability to analyze data, make informed decisions, and maintain integrity is essential. In my current role as a technical PM, I have had to make tough decisions and navigate complex projects, and I believe that my analytical skills and ability to stay true to my values have helped me to be successful in those situations. Like

Atticus, I strive to be a respected and trusted leader who can make a positive impact through my work.

Q2: If you could have dinner with any historical figure, who would it be and why?

Example response:

There are many historical figures who I would be interested in having dinner with, as each of them has made significant contributions to their respective fields and impacted the world in significant ways.

If I had to choose one figure, I might choose Albert Einstein. Einstein was a brilliant scientist and mathematician who made groundbreaking contributions to the fields of physics and mathematics, and his work has had a profound impact on our understanding of the world around us. I would be fascinated to hear Einstein's insights on the current state of scientific research and how he thinks his work has influenced the direction of science in the century since he made his discoveries. I would also be interested to hear his thoughts on the social and political issues of his time, as Einstein was known for his humanitarian efforts and his commitment to social justice. Overall, I believe that a dinner with Einstein would be a unique opportunity to learn from one of the most brilliant and influential figures in history.

Q3: If you could be any animal, what would you be and why?

Example response:

If I could be any animal, I might choose to be a dolphin. Dolphins are intelligent, social creatures who are known for their playful and curious nature. They are also highly adaptable and can be found in a wide range of habitats, from shallow coastal waters to open oceans.

I would be interested in being a dolphin because of their ability to communicate with one another and to work together as a group to achieve common goals. I would also enjoy being able to swim and explore the underwater world, and to have the opportunity to learn more about the behaviors and habits of other marine animals. Overall, I believe that being a dolphin would provide a unique and enriching experience, and I would be excited to learn more about the world from a dolphin's perspective.

Q4: If you could have any superpower, what would it be and why?

Example response:

If I could have any superpower, I would choose the ability to read people's thoughts and emotions. I believe that this would be a valuable skill for a product manager, as it would allow me to better understand the needs and motivations of my customers and stakeholders. By being able to read people's thoughts and emotions, I would be able to identify their pain points and unmet needs, and I could use this information to inform my product roadmap and development decisions.

In addition, this superpower would enable me to build stronger relationships with my team members and stakeholders, as I would be able to understand their perspectives and concerns more deeply. I believe that this would be particularly useful in a product manager role, where building strong relationships with cross-functional teams is critical to achieving success.

Overall, I believe that the ability to read people's thoughts and emotions would be a powerful tool for a product manager, and I would be excited to use it to build better products and stronger teams.

Q5: If you could travel back in time, what period would you go to and why?

Example response:

If I could travel back in time, I would choose to visit the Industrial Revolution. This was a transformative period in history that had a significant impact on the way we live and work today, and I would be interested in learning more about the technological and social changes that occurred during this time.

As a product manager, I believe that understanding the historical context and evolution of different industries is important for success in my role. By visiting the Industrial Revolution, I could learn more about the early development of modern manufacturing and transportation systems, and I could understand the ways in which these technologies have shaped the world we live in today.

In addition, I would be interested in exploring the cultural and social changes that occurred during the Industrial Revolution, and in understanding how these changes have impacted the way we live and work. I believe that this would be a valuable learning experience that would help me to understand the context and background of different industries and markets, and to make informed decisions as a product manager.

Q6: **If you could switch lives with any person for a day, who would it be and why?**

Example response:

If I could switch lives with any person for a day, I would choose to switch with Elon Musk. As a technical product manager, I am always looking for ways to learn and grow, and I believe that switching lives with someone who has achieved such incredible success in the field of technology and innovation would be a valuable learning experience.

Elon Musk has an incredible wealth of knowledge and experience in the field of technology and innovation, and I would be excited to have the opportunity to learn from him and to see the world through his eyes. He has a unique perspective on the intersection of technology and society, and I would be interested in understanding how he approaches and solves complex technical challenges.

By switching lives with Elon Musk for a day, I could gain insights into how he develops and implements innovative solutions, and I could see firsthand how he approaches problem-solving and decision-making. I could also learn more about how he builds and leads successful teams, and how he navigates the complex landscape of technology and innovation.

In addition to learning from Elon Musk's technical expertise and leadership skills, I would also be interested in understanding his perspective on the broader social and cultural implications of technology. This is an important consideration for a technical product manager, and I would be interested in understanding how Elon Musk approaches this aspect of his work.

Overall, I believe that switching lives with Elon Musk for a day would be an incredible opportunity to learn and grow as a technical product manager, and I would be excited to have the chance to experience the world from his perspective. I believe that this experience would provide me with valuable insights and skills that I could apply in my own work as a technical product manager, and I would be grateful for the opportunity to learn from such a successful and innovative leader.

Q7: If you could switch lives with any person for a day, who would it be and why?

Example response:

If I could have any job in the world, I would choose to be a product manager. Product management is an incredibly rewarding and challenging role that allows me to use my technical and business skills to solve complex problems and create value for customers and stakeholders.

As a technical product manager, I have the opportunity to work with a diverse group of people, including engineers, designers, and business professionals, and to lead cross-functional teams to deliver innovative and impactful products. This role requires a unique combination of technical expertise, business acumen, and leadership skills, and I enjoy the challenge of developing and growing these skills in my work.

One of the tools that I particularly enjoy using as a product manager is data analytics. Data is a critical resource for product managers, as it allows us to understand the needs and preferences of our customers and stakeholders, and to identify trends and patterns that can inform our product development and marketing strategies. I particularly enjoy using data visualization tools to identify insights and trends in large datasets, and this is a valuable skill for a product manager to have.

For example, in my previous role as a product manager for an online marketplace, I used data analytics and visualization tools to identify trends in customer behavior and preferences, and to identify opportunities for product innovation and optimization. This allowed me to develop targeted marketing campaigns and to optimize the user experience for our customers, resulting in increased engagement and revenue for the company.

In addition to my technical skills, I also enjoy the opportunity to work with customers and stakeholders to understand their needs and preferences, and to use this knowledge to create products that meet their needs and exceed their expectations. This is a critical part of the product manager role, and it is something that I am passionate about.

Overall, the product manager role is the perfect fit for me, and I am excited to continue learning and growing in this role. This role allows me to use my technical and business skills to make a meaningful impact on the lives of customers and stakeholders, and I am motivated by the opportunity to work with a diverse group of people to create innovative and impactful products.

Q8: If you could live in any city in the world, where would it be and why?

Example response:
If I could live in any city in the world, I would choose San Francisco. San Francisco is an incredible city that is at the forefront of innovation and technology, and I am drawn to its vibrant and diverse culture.

As a product manager, I believe that living in San Francisco would provide me with unique opportunities to learn and grow in my career. The city is home to many of the world's leading technology companies, and it has a rich ecosystem of startups and entrepreneurs that are driving innovation in a wide range of industries. I believe that living in San Francisco would allow me to be at the heart of this innovation, and to learn from and collaborate with some of the brightest minds in the field.

In addition to its thriving technology and innovation scene, San Francisco is also known for its rich cultural and artistic communities, which I believe would provide me with endless opportunities to learn and grow as a person. The city has a rich history of activism and social justice, and I believe that living in San Francisco would allow me to be a part of this vibrant and engaged community.

Q9: If you could be any fictional character, who would you be and why?

Example response:
If I could be any fictional character, I would choose to be Tony Stark from the Marvel Cinematic Universe. Tony Stark is an incredibly intelligent and resourceful character who can use his technical skills and creativity to solve complex problems and create innovative solutions.

As a technical product manager, these are important skills to have, and I am drawn to Tony Stark's ability to think outside the box and to use his technical expertise to create new and groundbreaking technologies. I also admire his ability to take risks and to challenge the status quo, which I believe are important qualities for a product manager to have.

In addition to his technical skills, Tony Stark is also a strong leader and a good communicator, which are important qualities for a product manager to have. He can inspire and motivate his team to achieve their goals, and he is able to effectively communicate his vision and ideas to stakeholders. These skills are critical for a product manager to have, and I am inspired by Tony Stark's ability to embody them in his work.

Q10: If you could have any hobby or pastime, what would it be and why?

Example response:

If I could have any hobby or pastime, I would choose to engage in competitive hacking events and challenges. This activity aligns with my technical skills and interests as a product manager, and it allows me to stay up to date on the latest technologies and best practices in cybersecurity.

I have a solid foundation in computer science and have experience with a variety of programming languages and frameworks. As a product manager, it is important to continue learning and growing in my technical skills and hacking provides an excellent opportunity to do so.

In addition to its technical benefits, I also find hacking to be an exciting and enjoyable hobby. I enjoy the sense of accomplishment that comes from successfully solving complex problems and finding creative solutions. I also appreciate the social aspect of hacking, as it allows me to connect with like-minded individuals and to learn from others in the community.

Overall, hacking is an ideal hobby for a technical product manager to have, as it allows me to continue learning and growing in my technical skills while also providing an exciting and challenging activity that I can enjoy in my free time.

Q11: If you could travel to any planet, which one would you choose and why?

Example response:

As a technical product manager, I would choose to travel to Mars because of the unique challenges and opportunities it presents in terms of exploration and research. Mars is a planet with a rich history and geology, and I am fascinated by the possibility of contributing to our understanding of these features through technical analysis and research.

There are several areas in which I believe my technical skills and knowledge could be applied to advance the exploration and understanding of Mars. For example, I have experience with systems engineering and robotics, and I believe that these skills would be valuable in designing and building systems and equipment for use on the planet. I am also interested in the challenges of establishing a human settlement on Mars, and I believe that my technical skills and knowledge could be applied to developing solutions for the various challenges that would need to be overcome to make this a reality.

In addition to the technical challenges and opportunities presented by Mars, I am also drawn to the sense of adventure and discovery that comes with exploring new frontiers. Mars represents a unique opportunity to contribute to the advancement of space exploration, and I am excited to have the opportunity to do so.

Q12: If you could be any age for the rest of your life, what age would you choose and why?

Example response:

If I could be any age for the rest of my life, I would choose to be around 35 years old. At this age, I believe that I have gained a good balance of life experience and technical knowledge, and I am at a point in my career where I am able to contribute significantly to projects and initiatives.

As a technical product manager, it is important to have a broad range of experiences and knowledge to draw upon to solve complex problems and to make informed decisions. I have spent the first half of my career building a strong foundation in technical skills and knowledge, and I am now able to use these skills effectively to contribute to projects and initiatives.

In addition to my technical skills and knowledge, I also believe that I have gained a good balance of life experience at this age. I have had the opportunity to travel, to work with diverse teams, and to learn from a variety of mentors and colleagues. I believe that these experiences have helped me to develop my communication and collaboration skills, and I am able to contribute significantly to teams and projects at this point in my career.

Q13: If you could live in any era, which one would you choose and why?

Example response:
If I could live in any era, I would choose to live in the present. As a technical product manager, the present offers the greatest opportunities to use my skills and knowledge to solve complex problems and to contribute to the advancement of technology and society.

One of the key reasons why I would choose to live in the present is the rapid pace of technology and innovation that we are currently experiencing. For example, we are seeing the development of technologies such as artificial intelligence, machine learning, and the Internet of Things, which have the potential to revolutionize industries and transform the way that we live and work. As a technical product manager, I am excited by the opportunity to work with these technologies and to contribute to their development and deployment.

In addition to the technical opportunities presented by the present, the present offers a wealth of data and insights that can be used to inform decision-making and to solve complex problems. With the proliferation of data analytics and visualization tools, we can collect and analyze vast amounts of data in real-time, and to use this data to inform our understanding of trends and patterns, and to identify opportunities for improvement. As a technical product manager, I am excited by the opportunity to work with data and to use it to inform my decision-making and problem-solving.

Q14: If you could have any talent or skill, what would it be and why?

Example response:

If I could have any talent or skill, I would choose the ability to speak and understand multiple languages fluently. As a technical product manager, I believe that the ability to communicate effectively with people from diverse cultures and backgrounds is an invaluable asset, and one that would enhance my ability to collaborate with diverse teams and to contribute to global projects and initiatives.

In today's globalized world, the ability to speak multiple languages allows you to communicate with people from around the world, and to understand their perspectives and needs. This is particularly important for product managers, who often work with cross-functional and international teams, and who need to be able to communicate effectively with people from a variety of cultures and backgrounds.

In addition to the communication benefits of speaking multiple languages, I also believe that this skill would provide me with a greater appreciation for diverse cultures and ways of life. This cultural awareness and understanding are important for product managers, as it allows them to be more sensitive to the needs and preferences of their customers and users, and to design products that are more tailored to those needs.

Overall, I believe that the ability to speak and understand multiple languages fluently would be a valuable talent for a technical product manager to have, and one that would enhance my ability to communicate effectively with diverse teams and to contribute to global projects and initiatives.

Q15: If you could be any fictional character's sidekick, who would you choose and why?

Example response:

As a product manager, I am drawn to the character of Sam because of his loyalty, his determination, and his resourcefulness. These are important qualities for a product manager to have, as they allow us to stay focused on our goals, to persevere through challenges, and to find creative solutions to problems. In addition to these qualities, I also bring a strong background in data analysis, user experience design, and project management to the role.

In my previous role as a product manager at XYZ company, I led the development and launch of a successful mobile app that was downloaded over one million times within the first year. I achieved this success by conducting extensive user research and analysis, developing a robust project plan, and managing a team of developers, designers, and QA analysts. I also utilized a range of technical tools and platforms, including JIRA, Asana, and Adobe Creative Suite, to ensure that the project was delivered on time and to the highest quality standards.

In addition to Sam's unwavering loyalty to Frodo, I also admire his ability to adapt and thrive in new and unfamiliar environments. As a product manager, I believe that the ability to adapt and learn quickly is critical to success, and that being open to new experiences and ideas is essential to driving innovation. I have demonstrated this ability in my previous roles, where I have successfully navigated the challenges of entering new markets and launching new products.

Overall, I believe that being Samwise Gamgee would be an incredible opportunity to work with and learn from a highly skilled and visionary leader, and to contribute to the development of innovative products and solutions. I am confident that my technical skills and determination would make me an asset to Frodo's team, and that I would thrive in this exciting and challenging role.

CHAPTER 8
Technical (Code) Based Questions Answers and Explanations

Q1: How would you design an algorithm to find the maximum element in a list of integers?

Example response:
1. Initialize a variable **max** to the first element in the list.
2. Iterate through the rest of the elements in the list. For each element, check if it is greater than the current value of **max**. If it is, update the value of **max** to the current element.
3. After the loop completes, **max** will contain the maximum element in the list.

Here is an example of how this algorithm might be implemented in Python:

```
def find_max(lst):
```

```
    max = lst[0]    # Initialize max to the first element in the list

    for i in range(1, len(lst)):    # Iterate through the rest of the elements

        if lst[i] > max:    # If the current element is greater than max

            max = lst[i]    # Update the value of max

    return max    # Return the maximum element
```

This algorithm has a time complexity of O(n), where n is the length of the list, because it requires one pass through the list to find the maximum element.

This answer provides a simple and straightforward algorithm for finding the maximum element in a list of integers. It is a good solution because it has a time complexity of O(n), which means that it will be efficient for lists of any size.

There are a few ways in which this solution could be improved:

1. If the list is already sorted in ascending order, you can stop the search as soon as you find an element that is less than the maximum element seen so far. This would reduce the time complexity to O(k), where k is the position of the maximum element in the list.

2. If the list is exceptionally large, you could use a divide and conquer approach to find the maximum element more efficiently. For example, you could split the list into two halves, find the maximum element in each half, and then return the larger of the two maximum elements. This would have a time complexity of O(log n).

Q2: How would you design an algorithm to sort a list of integers in ascending order?

Example response:

There are many different algorithms that can be used to sort a list of integers in ascending order. Here is one algorithm that you could use:

1. Initialize a variable **sorted** to an empty list.
2. While the list is not empty:
   - Find the minimum element in the list.
   - Remove the minimum element from the list.
   - Append the minimum element to the end of **sorted**.
3. Return **sorted** as the sorted list.

Here is an example of how this algorithm might be implemented in Python:

```
def sort_list(lst):

    sorted = []   # Initialize an empty list to hold the sorted elements

    while lst:    # While the list is not empty

        min_val = min(lst)   # Find the minimum element in the list

        lst.remove(min_val)  # Remove the minimum element from the list

        sorted.append(min_val)  # Append the minimum element to the end

    return sorted  # Return the sorted list
```

This algorithm has a time complexity of $O(n^2)$, where n is the length of the list, because it requires one pass through the list to find the minimum element and another pass to remove it. It also has a space complexity of O(n) because it creates a new list to hold the sorted elements.

While this is a good solution for small lists because it is easy to understand and implement, this algorithm has a time complexity of O(n^2), which means that it may not be efficient for larger lists. There are many other algorithms that can be used to sort a list of integers more efficiently, including quicksort and merge sort, which have time complexities of O(n log n) in the average case. If you are looking to improve the performance of this algorithm, you could consider using a more efficient sorting algorithm such as quicksort or merge sort. You could also consider using a parallel algorithm, which would allow you to take advantage of multiple processors or cores to sort the list more quickly.

Q3: How would you design an algorithm to find the shortest path between two nodes in a graph?

Example response:
To design an algorithm to find the shortest path between two nodes in a graph, you can use a graph traversal algorithm such as breadth-first search (BFS).

BFS is an algorithm that starts at a given node and explores all its neighboring nodes before moving on to the next level of nodes. It continues this process until it reaches the destination node or determines that there is no path from the starting node to the destination node.

Here is an example of how BFS might be implemented in Python:

```
def shortest_path(graph, start, end):
    # Initialize a queue to hold the nodes to be visited
    queue = [(start, [start])]
    # Initialize a set to keep track of visited nodes
    visited = set()

    while queue:
        # Dequeue the next node to be visited
        (node, path) = queue.pop(0)
        # If the node has not been visited
        if node not in visited:
            # Mark the node as visited
            visited.add(node)
            # If the node is the destination node, return the path
            if node == end:
                return path
            # Enqueue the neighbors of the node
            for neighbor in graph[node]:
                queue.append((neighbor, path + [neighbor]))
    # If there is no path from the start to the end, return None
    return None
```

This algorithm has a time complexity of O(V + E), where V is the number of nodes in the graph and E is the number of edges, because it requires one pass through the graph to find the shortest path. It has a space complexity of O(V), because it stores the nodes in the queue and visited set.

BFS is a good algorithm to use for finding the shortest path between two nodes in a graph because it guarantees that the first path it finds is the shortest path. However, it may not be the most efficient algorithm for larger graphs, as it requires storing the entire path in memory. In such cases, you may want to consider using a more efficient algorithm such as Dijkstra's algorithm or A* search.

Q4: How would you design an algorithm to implement a stack data structure?

Example response:

To design an algorithm to implement a stack data structure, you can use the following steps:

1. Define a class **Stack** with the following methods:
   - **__init__**: Initializes an empty stack.
   - **push(item)**: Adds an item to the top of the stack.
   - **pop()**: Removes and returns the item at the top of the stack.
   - **peek()**: Returns the item at the top of the stack without removing it.
   - **is_empty()**: Returns **True** if the stack is empty, **False** otherwise.

2. Implement the **Stack** class using a list to store the items in the stack.
   - To add an item to the stack, use the **append** method to add the item to the end of the list.

- To remove an item from the stack, use the **pop** method to remove the item from the end of the list.
- To return the item at the top of the stack, use the **-1** index to access the last element in the list.
- To check if the stack is empty, use the **len** function to check the length of the list.

Here is an example of how the **Stack** class might be implemented in Python:

```
class Stack:

    def __init__(self):

        self.items = []  # Initialize an empty list to store the items in the stack


    def push(self, item):

        self.items.append(item)  # Add the item to the end of the list


    def pop(self):

        return self.items.pop()  # Remove and return the item from the end of the list


    def peek(self):

        return self.items[-1]  # Return the item at the end of the list without removing it


    def is_empty(self):

        return len(self.items) == 0  # Return True if the list is empty, False otherwise
```

This implementation of the **Stack** class has a time complexity of O(1) for the **push**, **pop**, and **peek** methods, because they all operate on the end of the list, which has a constant time complexity. The **is_empty** method has a time complexity of O(1) because it simply checks the length of the list. The space complexity is O(n), where n is the number of items in the stack, because it stores the items in a list.

There are a few ways in which this solution could be improved:

1. If the stack will be used to store a large number of items, you could consider using a more efficient data structure such as a linked list to store the items. This would improve the space complexity, as a linked list uses less space than a list for large collections of items.

2. You could implement the **Stack** class using a custom linked list data structure rather than a built-in list. This would allow you to optimize the implementation for the specific requirements of the stack data structure.

3. You could add additional methods to the **Stack** class to provide additional functionality, such as a **size** method to return the number of items in the stack or a **clear** method to remove all items from the stack.

Q5: How would you design an algorithm to implement a queue data structure?

Example response:
1. Define a class **Queue** with the following methods:
   - **__init__**: Initializes an empty queue.
   - **enqueue(item)**: Adds an item to the end of the queue.

- **dequeue()**: Removes and returns the item at the front of the queue.
- **peek()**: Returns the item at the front of the queue without removing it.
- **is_empty()**: Returns **True** if the queue is empty, **False** otherwise.

2. Implement the **Queue** class using a list to store the items in the queue.

- To add an item to the queue, use the **append** method to add the item to the end of the list.
- To remove an item from the queue, use the **pop** method to remove the item from the front of the list.
- To return the item at the front of the queue, use the **0** index to access the first element in the list.
- To check if the queue is empty, use the **len** function to check the length of the list.

Here is an example of how the **Queue** class might be implemented in Python:

```
class Queue:
    def __init__(self):
        self.items = []  # Initialize an empty list to store the items in the queue

    def enqueue(self, item):
        self.items.append(item)  # Add the item to the end of the list

    def dequeue(self):
```

```
            return self.items.pop(0)   # Remove and return the
    item from the front of the list


        def peek(self):

            return self.items[0]   # Return the item at the front
    of the list without removing it


        def is_empty(self):

            return len(self.items) == 0   # Return True if the
    list is empty, False otherwise
```

This implementation of the **Queue** class has a time complexity of O(1) for the **enqueue** and **peek** methods, because they operate on the end of the list, which has a constant time complexity. The **dequeue** method has a time complexity of O(n) because it requires shifting all the other items in the list down by one index. The **is_empty** method has a time complexity of O(1) because it simply checks the length of the list. The space complexity is O(n), where n is the number of items in the queue, because it stores the items in a list.

There are many other ways to implement a queue data structure, including using a linked list or a circular buffer. The choice of implementation will depend on the specific requirements and constraints of the problem.

Q6: How would you design an algorithm to reverse a string?

Example response:

To design an algorithm to reverse a string, you can use the following steps:

1. Define a function **reverse(s)** that takes in a string **s** as input and returns the reversed string.
2. Initialize an empty string **reversed** to hold the reversed string.
3. Iterate over the string **s** in reverse order, starting from the last character and ending at the first character. For each character **c**, do the following:

- Append **c** to the end of **reversed**.

4. Return **reversed** as the reversed string.

Here is an example of how this algorithm might be implemented in Python:

```python
def reverse(s):
    reversed = ""   # Initialize an empty string to hold the reversed string

    # Iterate over the string in reverse order
    for i in range(len(s)-1, -1, -1):
        # Append the character to the end of reversed
        reversed += s[i]

    return reversed  # Return the reversed string
```

This algorithm has a time complexity of $O(n)$, where n is the length of the string, because it requires one pass through the string to reverse it. It has a space complexity of $O(n)$ because it creates a new string to hold the reversed string.

Q7: How would you design an algorithm to find the least common multiple of two integers?

Example response:

To design an algorithm to find the least common multiple (LCM) of two integers, you can use the following steps:

1. Define a function **lcm(a, b)** that takes in two integers **a** and **b** as input and returns their LCM.
2. Initialize a variable **lcm** to the maximum of **a** and **b**.
3. Implement a loop to iterate over the range **[lcm, a*b + 1]**. For each number **i** in the range, do the following:

   - If **i % a == 0** and **i % b == 0**, return **i** as the LCM.

4. If the loop completes without finding a common multiple, return **-1** as the LCM.

Here is an example of how this algorithm might be implemented in Python:

```
def lcm(a, b):
    # Initialize lcm to the maximum of a and b
    lcm = max(a, b)
    # Iterate over the range [lcm, a*b + 1]
    for i in range(lcm, a*b + 1):
        # If i is a common multiple of a and b, return i as the lcm
        if i % a == 0 and i % b == 0:
            return i
    # If the loop completes without finding a common multiple, return -1 as the lcm
```

```
return -1
```

This algorithm has a time complexity of O(n), where n is the maximum of **a** and **b**, because it iterates over a range of numbers up to the maximum of **a** and **b**. It has a space complexity of O(1), because it only uses a constant amount of memory to store the variables **lcm** and **i**.

Q8: How would you design an algorithm to find the factorial of a given number?

Example response:

To design an algorithm to find the factorial of a given number, you can use the following steps:

1. Define a function **factorial(n)** that takes in an integer **n** as input and returns the factorial of **n**.

2. Implement a loop to iterate over the range **[1, n+1]**. For each number **i** in the range, do the following:

   - Initialize a variable **result** to 1.

   - Multiply **result** by **i** and assign the result back to **result**.

3. Return **result** as the factorial of **n**.

Here is an example of how this algorithm might be implemented in Python:

```
def factorial(n):
    result = 1  # Initialize result to 1
    # Iterate over the range [1, n+1]
    for i in range(1, n+1):
        # Multiply result by i and assign the result back
```

```
            to result
                result *= i
            return result  # Return the result as the factorial of n
```

This algorithm has a time complexity of O(n), where n is the input number, because it iterates over a range of numbers up to **n**. It has a space complexity of O(1), because it only uses a constant amount of memory to store the variables **i** and **result**.

Q9: How would you design an algorithm to check if a given string is a palindrome?

Example response:

To design an algorithm to check if a given string is a palindrome, you can use the following steps:

1. Define a function **is_palindrome(s)** that takes in a string **s** as input and returns a boolean indicating whether the string is a palindrome.

2. Implement a loop to iterate over the string **s**, starting from the first character and ending at the middle character. For each character **c** at index **i**, do the following:

   - If **c** is not equal to the character at index **len(s) - i - 1**, return **False** as the result.

3. If the loop completes without finding any mismatched characters, return **True** as the result.

Here is an example of how this algorithm might be implemented in Python:

```
def is_palindrome(s):

    # Iterate over the string, starting from the first char-
    acter and ending at the middle character

    for i in range(len(s)//2):

        # If the characters at index i and len(s) - i - 1
        are not equal, return False

        if s[i] != s[len(s) - i - 1]:

            return False

    # If the loop completes without finding any mismatched
    characters, return True

    return True
```

This algorithm has a time complexity of O(n), where n is the length of the string, because it requires one pass through the string to check if it is a palindrome. It has a space complexity of O(1), because it only uses a constant amount of memory to store the variables **i** and **c**.

Q10: How would you design an algorithm to find the nth Fibonacci number?

Example response:

To find the nth Fibonacci number, you can use the following function:

```
def fibonacci(n):
    if n == 0:
        return 0
    elif n == 1:
        return 1
    else:
        return fibonacci(n-1) + fibonacci(n-2)
```

This function uses recursion to compute the nth Fibonacci number. It has a base case for when **n** is 0 or 1, in which it returns 0 or 1, respectively. For all other values of **n**, it returns the sum of the previous two Fibonacci numbers.

This function has a time complexity of O(2^n) because it makes two recursive calls for each call to **fibonacci**. It has a space complexity of O(n) because it stores the recursive calls in the call stack.

There are a few ways to improve the performance of the function to find the nth Fibonacci number:

1. *Use dynamic programming:* Instead of computing the Fibonacci numbers from scratch every time, you can use an array to store the computed values and reuse them as needed. This will reduce the time complexity to O(n) and the space complexity to O(n).
2. *Use a loop:* Instead of using recursion, you can use a loop to iterate over the range **[2, n]** and compute the Fibonacci numbers using a bottom-up approach. This will reduce the time complexity to O(n) and the space complexity to O(1).
3. *Use a closed-form formula:* There are several closed-form formulas that can be used to compute the nth Fibonacci number in constant time and constant space. One such formula is **F(n) = ((1 + sqrt(5))/2)^n / sqrt(5)**, where **sqrt** is the square root function.
4. *Use a matrix exponentiation algorithm:* You can use matrix exponentiation to compute the nth Fibonacci number in O(log n) time and O(1) space.

CHAPTER 9

Scenario Based Technical (Code) Questions Answers and Explanations

Q1: Your company has a large database of customer records. How would you design an algorithm to find the customer with the highest lifetime value?

Example response:
1. Define a function **find_customer_with_highest_ltv(customers)** that takes in a list of customer records **customers** as input and returns the customer with the highest lifetime value.
2. Initialize a variable **max_ltv** to the minimum integer value and a variable **max_customer** to **None**.
3. Iterate over the **customers** list. For each customer **c**, do the following:
   - Calculate the lifetime value of **c** by summing the values of all their purchases.

- If the lifetime value of **c** is greater than **max_ltv**, set **max_ltv** to the lifetime value of **c** and **max_customer** to **c**.

4. Return **max_customer** as the customer with the highest lifetime value.

This algorithm has a time complexity of O(n), where n is the number of customers, because it requires one pass through the list of customers to find the customer with the highest lifetime value. It has a space complexity of O(1), because it only uses a constant amount of memory to store the variables `max_ltv` and `max_customer`.

There are a few ways to improve the performance of the algorithm to find the customer with the highest lifetime value:

1. *Use a more efficient data structure:* Instead of using a list to store the customer records, you could use a data structure that allows for faster access to the customer records. For example, you could use a hash table or a tree data structure to store the customer records, which would allow you to look up a customer's record in O(1) or O(log n) time, respectively.

2. *Use a more efficient sorting algorithm:* If you choose to use a sorting algorithm to find the customer with the highest lifetime value, you could use a more efficient sorting algorithm such as quicksort or merge sort, which have a time complexity of O(n log n) in the average case.

3. *Use parallel processing:* If you have access to multiple processors or cores, you could use parallel processing to speed up the calculation of the lifetime values for each customer. This would allow you to calculate the lifetime values for multiple

customers simultaneously, which would reduce the overall time complexity of the algorithm.

4. *Use an approximate solution:* If the exact solution is not required, you could use an approximate solution such as sampling a subset of the customer records and selecting the customer with the highest lifetime value from the sample. This would allow you to find the customer with the highest lifetime value in $O(k)$ time, where k is the size of the sample, at the cost of potentially returning an approximate result.

Q2: Your company has a large database of customer records. How would you design an algorithm to find the customer with the highest lifetime value?

Example response:
To design an algorithm to select the appropriate customers based on their purchase history and preferences, you can use the following steps:

1. Define a function **select_customers(customers, product)** that takes in a list of customer records **customers** and a product **product** as input and returns a list of customers who should be targeted for promotional emails.
2. Initialize a list **selected_customers** to store the selected customers.
3. Iterate over the **customers** list. For each customer **c**, do the following:
   - If **c** has purchased a similar product in the past, add **c** to the **selected_customers** list.

- If **c** has expressed interest in the product category that **product** belongs to, add **c** to the **selected_customers** list.

4. Return **selected_customers** as the list of customers who should be targeted for promotional emails.

There are many other ways to select the appropriate customers based on their purchase history and preferences, depending on the specific requirements and constraints of the problem. For example, you could use a more sophisticated scoring system to prioritize customers based on their likelihood to purchase the product, or you could use machine learning techniques to predict which customers are most likely to be interested in the product.

Q3: Your company is developing a recommendation engine for a streaming service. How would you design an algorithm to recommend similar items to a user based on their past viewing history?

Example response:

To design an algorithm to recommend similar items to a user based on their past viewing history, you can use the following steps:

1. Define a function **recommend_similar_items(user, items)** that takes in a user **user** and a list of items **items** as input and returns a list of recommended items.
2. Calculate the similarity between each item in the **items** list and the items in the user's viewing history. This can be done using a similarity measure such as cosine similarity or Pearson correlation coefficient.

3. Sort the items in the **items** list in descending order of similarity to the items in the user's viewing history.

4. Return the top N items from the sorted list as the recommended items, where N is the number of recommendations desired.

This algorithm has a time complexity of O(n log n), where n is the number of items, because it requires one pass through the list of items to calculate the similarities and another pass to sort the items. It has a space complexity of O(n) because it stores the items in a list.

There are many other ways to recommend similar items to a user based on their past viewing history, depending on the specific requirements and constraints of the problem. For example, you could use a more sophisticated recommendation algorithm such as collaborative filtering or matrix factorization, or you could use machine learning techniques to build a recommendation model.

Q4: Your company is building a chatbot to assist customers with their orders. How would you design an algorithm to handle multiple concurrent conversations and route them to the appropriate customer service representative?

Example response:
To design an algorithm to handle multiple concurrent conversations and route them to the appropriate customer service representative, you can use the following steps:

1. Define a function **route_conversation(conversation, customer_service_representatives)** that takes in a conversation **conversation** and a list of customer service representatives **customer_service_representatives** as input and returns

the customer service representative who should handle the conversation.

2. Calculate the workload of each customer service representative by summing the number of conversations they are currently handling.

3. Sort the customer service representatives in ascending order of workload.

4. Return the customer service representative with the lowest workload as the one who should handle the conversation.

This algorithm has a time complexity of O(n log n), where n is the number of customer service representatives, because it requires one pass through the list of customer service representatives to calculate the workloads and another pass to sort the customer service representatives. It has a space complexity of O(n) because it stores the customer service representatives in a list.

There are various approaches that can be taken to manage multiple concurrent conversations and assign them to the most suitable customer service representative, based on the specific needs and constraints of the situation. For instance, you might employ a more advanced routing strategy like round-robin scheduling or load balancing, or you could use machine learning techniques to forecast the workload of each customer service representative and direct conversations accordingly.

Q5: Your company is building a ride-sharing app and wants to optimize the matching of riders with drivers. How would you

design an algorithm to find the best match based on location, availability, and ratings?

Example response:

To design an algorithm to find the best match between riders and drivers based on location, availability, and ratings, you can use the following steps:

1. Define a function **find_best_match(rider, drivers)** that takes in a rider **rider** and a list of drivers **drivers** as input and returns the best match between the rider and the drivers.

2. Filter the list of drivers to only include those who are within a certain distance of the rider's pickup location and are available to accept a ride.

3. Sort the list of drivers in ascending order of distance from the rider's pickup location.

4. For each driver **d** in the sorted list of drivers, do the following:

   - Calculate the match score between the rider and the driver by taking the average of the rider's rating, the driver's rating, and the distance between the rider's pickup location and the driver's current location.

   - If the match score is above a certain threshold, return the driver as the best match.

5. If no driver is found to be a good match, return **None** as the best match.

This algorithm has a time complexity of $O(n)$, where n is the number of drivers, because it requires one pass through the list of drivers to filter and sort the drivers and another pass to calculate the match scores. It has a space complexity of O(n) because it stores the drivers in a list.

Other methods for finding the best match between riders and drivers based on location, availability, and ratings could be considered depending on the specific requirements and constraints of the situation. Examples of these methods include using a more advanced matching algorithm such as linear programming or graph matching, or using machine learning techniques to predict the success rate of a match between a rider and a driver.

Q6: Your company is building a language translation app and wants to optimize the translation process. How would you design an algorithm to choose the best translation for a given word or phrase based on context and past translations?

Example response:

To design an algorithm to choose the best translation for a given word or phrase based on context and past translations, you can use the following steps:

1. Define a function **choose_best_translation(word_or_phrase, translations, context)** that takes in a word or phrase **word_or_phrase**, a list of translations **translations**, and context information **context** as input and returns the best translation for the word or phrase.
2. Use the context information to filter the list of translations to only include those that are relevant to the current context.
3. Sort the list of translations in descending order of frequency of use in past translations.

4. Return the top translation from the sorted list as the best translation.

This algorithm has a time complexity of $O(n \log n)$, where n is the number of translations, because it requires one pass through the list of translations to filter the translations and another pass to sort the translations. It has a space complexity of $O(n)$ because it stores the translations in a list.

There are various approaches that can be taken to select the most accurate translation for a specific word or phrase based on context and previous translations, depending on the specific needs and constraints of the situation. These approaches might include using a more advanced translation algorithm like machine translation or neural machine translation or applying machine learning techniques to create a translation model.

Q7: **Your company is developing a fraud detection system for online transactions. How would you design an algorithm to identify suspicious activity and flag it for further review?**

Example response:
1. Define a function **detect_suspicious_activity(transaction)** that takes in a transaction **transaction** as input and returns **True** if the transaction is suspicious and **False** otherwise.
2. Extract features from the transaction such as the amount, the merchant, the location, and the payment method.
3. Use the extracted features to calculate a suspiciousness score for the transaction. This can be done using a machine learning

model trained on a dataset of labeled transactions or using a set of rules based on known fraudulent behavior.

4. If the suspiciousness score is above a certain threshold, return **True** to flag the transaction for further review. Otherwise, return **False**.

This algorithm has a time complexity of O(1), because it only requires a single pass through the transaction to extract the features and calculate the suspiciousness score. It has a space complexity of O(1), because it does not require any additional storage beyond the transaction itself.

Q8: Your company is building a virtual personal assistant to help users manage their schedules and tasks. How would you design an algorithm to understand and interpret user requests and provide appropriate responses?

Example response:
To design an algorithm to understand and interpret user requests and provide appropriate responses in a virtual personal assistant, you can use the following steps:

1. Define a function **process_request(request)** that takes in a user request **request** as input and returns a response.
2. Use natural language processing (NLP) techniques to parse the request and extract the relevant information such as the task, the deadline, and any other details.
3. Use a task management system or database to store and organize the tasks and deadlines.
4. Use a decision tree or similar machine learning model to classify the request and determine the appropriate action to take.

5. Return a response to the user based on the action taken.

This algorithm has a time complexity of O(1), because it only requires a single pass through the request to parse and classify it. It has a space complexity of O(1), because it does not require any additional storage beyond the request itself.

There are several ways to improve this answer:

1. Add more details about the specific NLP techniques and machine learning models that could be used to parse and classify the requests.
2. Provide more examples of different actions that the virtual personal assistant might take based on the user requests.
3. Discuss how the virtual personal assistant could handle ambiguous or incorrect requests, or requests that it is not programmed to handle.
4. Explain how the virtual personal assistant could learn and improve over time based on user feedback and usage data.
5. Describe how the virtual personal assistant could integrate with other systems or services to provide more comprehensive assistance to the user.
6. Discuss the potential security and privacy concerns that could arise when building a virtual personal assistant, and how they could be addressed.

Q9: Your company is building a customer support platform and wants to optimize the routing of incoming inquiries to the appropriate team or agent. How would you design

an algorithm to classify and route inquiries based on topic and complexity?

Example response:

To design an algorithm to classify and route incoming inquiries to the appropriate team or agent based on topic and complexity in a customer support platform, you can use the following steps:

1. Define a function **route_inquiry(inquiry)** that takes in an inquiry **inquiry** as input and returns the appropriate team or agent to handle the inquiry.
2. Use natural language processing (NLP) techniques to extract the relevant information from the inquiry such as the topic, the complexity, and any other details.
3. Use a decision tree or similar machine learning model to classify the inquiry based on the extracted information.
4. Use a routing table or database to determine the appropriate team or agent to handle the inquiry based on the classification.
5. Return the team or agent to the caller.

Q10: Your company is building an e-commerce platform and wants to optimize the search functionality for users. How would you design an algorithm to return the most relevant search results based on user queries and past searches?

Example response:

To design an algorithm to return the most relevant search results based on user queries and past searches in an e-commerce platform, you can use the following steps:

1. Define a function **search(query)** that takes in a user query **query** as input and returns a list of relevant search results.
2. Use natural language processing (NLP) techniques to preprocess the query and extract the relevant keywords.
3. Use a search index or database to retrieve the relevant search results based on the keywords.
4. Use a ranking algorithm such as TF-IDF or PageRank to score the search results based on their relevance to the query and their popularity among users.
5. Return the top-ranked search results to the user.

This algorithm has a time complexity of O(n), where n is the number of search results retrieved from the search index or database. It has a space complexity of O(n) because it requires storage for the search results and the ranked results.

CHAPTER 10

Brainteaser Questions Answers and Explanations

Q1: Your company is building an e-commerce platform and wants to optimize the search functionality for users. How would you design an algorithm to return the most relevant search results based on user queries and past searches?

Example response:
To find the ball that is different using the balance scale only twice, you can follow these steps:

1. Divide the 10 balls into two groups of 5 balls each.
2. Place the two groups of balls on the balance scale. If one group is heavier, the ball that is different is in that group. If the two groups are equal, the ball that is different is not in either group.
3. If one group is heavier, divide that group into two smaller groups of 2 balls and 3 balls.

4. Place the two smaller groups on the balance scale. If one group is heavier, the ball that is different is in that group. If the two groups are equal, the ball that is different is in the other group.

5. If one group is heavier, divide that group into two smaller groups of 1 ball each.

6. Place the two smaller groups on the balance scale. The ball that is different will be in the group that is heavier.

Q2: You are given two boxes, one containing only apples and the other containing only oranges. The boxes are labeled incorrectly, and you cannot see inside them. How can you determine which box contains the apples and which box contains the oranges using the scale only twice?

Example response:

To determine which box contains the apples and which box contains the oranges using the scale only twice, you can follow these steps:

1. Take one fruit from box A and one fruit from box B and place them on the scale. If one fruit is heavier, the heavier fruit is an apple. If the two fruits are equal, they are either both apples or both oranges.

2. If one fruit is heavier, the box containing the heavier fruit contains the apples. If the two fruits are equal, take one more fruit from each box and place them on the scale. If one fruit is heavier, the heavier fruit is an apple. If the two fruits are equal, they are both apples.

3. If one fruit is heavier, the box containing the heavier fruit contains the apples. If the two fruits are equal, the boxes contain apples and oranges.

This algorithm has a time complexity of O(1), because it only requires two passes through the fruits to determine which box contains the apples and which box contains the oranges. It has a space complexity of O(1), because it does not require any additional storage beyond the fruits themselves.

Q3: You are given a stack of cards, each with a number on one side and a letter on the other side. You are allowed to flip over two cards at a time. How would you determine which cards have the same number or letter, using the least number of flips?

Example response:
To determine which cards have the same number or letter using the least number of flips, you can follow these steps:

1. Divide the stack of cards into two piles: one pile containing the cards with numbers on one side and letters on the other side, and the other pile containing the cards with letters on one side and numbers on the other side.
2. Flip over one card from each pile and place them on the table. If the two cards have the same number or letter, you have found a pair. If the two cards do not have the same number or letter, return them to their respective piles.
3. Repeat step 2 until you have found all the pairs.

This algorithm has a time complexity of O(n), where n is the number of cards in the stack. It has a space complexity of O(1), because it does not require any additional storage beyond the cards themselves.

Q4: You are given two ropes and a lighter. Each rope takes exactly one hour to burn, but they do not burn at a consistent rate. One rope may burn faster or slower than the other. How would you measure out 45 minutes using only these two ropes and the lighter?

Example response:

To measure 45 minutes using the two ropes and a lighter, you can follow these steps:

1. Light one of the ropes at both ends and set it aside.
2. Light the other rope at one end and start a timer.
3. When the second rope has burned for 30 minutes, extinguish the flame at the unburnt end.
4. Wait for the first rope that you set aside to burn completely (this should take about 30 minutes).
5. At the exact moment that the first rope finishes burning, light the second rope at the other end.
6. Wait for the second rope to burn completely (this should take about 15 minutes).

This process will take a total of 45 minutes. The first rope burning for 30 minutes and the second rope burning for 15 minutes will add up to 45 minutes.

Q5: You are given a pile of marbles, some of which are red and some of which are green. You are allowed to pick up two marbles at a time. How would you determine the color of all the marbles in the least number of picks?

Example response:

To determine the color of all the marbles in the pile in the least number of picks, you can follow this strategy:

1. Pick up two marbles at a time and compare their colors.
2. If the two marbles are the same color, continue picking up two marbles at a time and comparing their colors.
3. If the two marbles are different colors, set one of the marbles aside and pick up another marble.
4. Compare the color of the new marble to the color of the marble that you set aside. If they are the same, then all the marbles in the pile are that color. If they are different, then there are two colors of marbles in the pile.

This process will allow you to determine the color of all the marbles in the pile in just three picks. If all the marbles are the same color, you will need only two picks to determine their color. If there are two colors of marbles in the pile, you will need three picks.

Q6: You are given a set of three light bulbs and three switches. Each switch controls one of the light bulbs, but you do not know which switch controls which light bulb. How would you determine the correct switch for each light bulb in the fewest number of tries?

Example response:

To determine the correct switch for each light bulb in the fewest number of tries, you can follow this strategy:

1. Turn on switches 1 and 2.
2. Wait a moment, then turn off switch 2.
3. Enter the room and observe which light bulbs are on and which are off.
4. The light bulb that is on is controlled by switch 1. The light bulb that is off and hot (i.e., still warm to the touch) is controlled by switch 2. The remaining light bulb is controlled by switch 3.

This process will allow you to determine the correct switch for each light bulb in just two tries. In the first try, you will turn on switches 1 and 2 and observe which light bulbs are on and which are off. In the second try, you will turn off switch 2 and observe which light bulb is now off and hot. After two tries, you will have identified the correct switch for each light bulb.

CHAPTER 11

Encouragement

First, congratulations on making it to the interview stage! This is a huge accomplishment and a testament to your hard work and dedication. The PM interview can be intimidating, but it is also a fantastic opportunity to highlight your skills and experience.

Here are a few tips to help you prepare:

1. Research the company and the role. Make sure you have a good understanding of the company's products, services, and culture. Also, take some time to review the job description and think about how your skills and experience align with the role.

2. Practice your communication skills. As a PM, you will need to be able to communicate your ideas and plans clearly and effectively. Practice your presentation skills by presenting to friends or colleagues, or even just talking through your thoughts aloud.

3. Prepare for common PM interview questions.

4. Think about your long-term goals. The PM interview is also an opportunity for you to share your career aspirations and discuss how this role fits into your overall goals.

5. Relax and be yourself. It is natural to feel nervous before an interview but try to stay calm and focused. Remember, the interviewer is looking for the best fit for the role, and that means finding someone who is authentic and genuine.

Finally, remember that the PM interview is just one step in the process. Do not put too much pressure on yourself – focus on doing your best and showing off your skills and experience. If this opportunity is not the right fit for you, there will be other opportunities in the future.

Good luck, you have this!

www.ingramcontent.com/pod-product-compliance
Lightning Source LLC
Chambersburg PA
CBHW071402210526
45465CB00001B/208